Fluid-Structure Interactions and Uncertainties

Reliability of Multiphysical Systems Set

coordinated by
Abdelkhalak El Hami

Volume 6

Fluid-Structure Interactions and Uncertainties

Ansys and Fluent Tools

Abdelkhalak El Hami
Bouchaib Radi

iSTE

WILEY

First published 2017 in Great Britain and the United States by ISTE Ltd and John Wiley & Sons, Inc.

ISTE Ltd
27-37 St George's Road
London SW19 4EU
UK

www.iste.co.uk

John Wiley & Sons, Inc.
111 River Street
Hoboken, NJ 07030
USA

www.wiley.com

Library of Congress Control Number: 2016960066

British Library Cataloguing-in-Publication Data
A CIP record for this book is available from the British Library
ISBN 978-1-84821-939-7

MIX
Paper from responsible sources
FSC® C013604

Contents

Preface

The progress achieved by digital and software tools in the past 40 years has allowed scientists to dramatically improve their understanding of the world. The development of mathematical models has allowed us to work on increasingly sophisticated problems in a wide range of fields: predicting the behavior of production tools, transportation, the environment, etc. Managing these complex problems has been facilitated within each discipline separately but also from a cross-disciplinary perspective, allowing more general phenomena to be tackled.

The field of fluid–structure interactions unites the study of all phenomena involving a coupling between the motion of a structure and the motion of a fluid. The range of studied phenomena is very broad, from vibrating cylinders in flows, such as in the nuclear industry, to vibrating structures in turbulent flows and free-surface phenomena in reservoirs. One well-known example of fluid–structure interaction and the complexity of the couplings involved is the collapse of the Tacoma bridge in 1940, which began vibrating to the point of resonance frequency under the effect of violent winds, causing it to be completely destroyed. This shows just how important it is to prepare reliable models in advance of any project so that this kind of behavior may be predicted.

Wind tunnels, such as in aeronautics, allow us to inspect the behavior of the structure on the ground without needing to perform tests in flight. The Euler and Navier–Stokes equations have made it possible to rigorously define a physical framework for characterizing the behavior of the aircraft in terms of a set of different parameters such as the velocity or Mach number. Finite element models have greatly simplified the process of representing an aircraft model and its structure, as well as the way that the aircraft responds to stress.

However, the complexity of the studied phenomena is reflected in the prohibitive computational costs, which motivates us to search for reduced models with more realistic computation times. By a reduced model, we mean a description as a low-dimensional system obtained by analyzing classical numerical formulations. Acheiving this reduction incurs an initial cost, but this cost is largely offset if the reduced model is later found to be applicable for configurations of parameters other than those of the initial formulation.

Thus, just like in other areas of the industry, optimization research is extremely active within the aviation sector. One significant development since the late 1980s has been the introduction of uncertainty parameters into numerical models. Optimization techniques in the presence of uncertainty in aerodynamics have only been studied more recently, beginning in the early 2000s. Their introduction was motivated by the need to account for specific types of situations that make it too difficult to precisely evaluate the aircraft's behavior. For example, during the aircraft design phase, in order to meet the various different criteria or eliminate certain problems encountered by the model, the model is able to adjust itself to more effectively meet the requirements and needs that it is designed to satisfy. The initial drafts of the model are not fixed, but for safety reasons it is necessary to ensure throughout the development process that the structure is capable of withstanding the stresses that it is likely to encounter in operating conditions. One way of accounting for these potential changes is to introduce uncertainty into the model.

Furthermore, when designing aircraft, manufacturers are naturally interested in maximizing the performance of each vehicle: reducing pollution, noise, drag, increasing the range, maximizing stability, etc. Minimizing the structural mass is an important objective for manufacturers as it allows other optimization criteria such as reducing pollution or extending the range to be satisfied. But less mass will also have negative repercussions on other criteria, including the stability of the aircraft in flight, for example by rendering it susceptible to the phenomenon of "fluttering".

Manufacturers must therefore perform constrained optimization: minimizing the weight of the wing while ensuring that fluttering cannot affect the airplane within its flight envelope. In such a case, optimization problems have a cross-disciplinary character, since they exhibit behaviors that include both structural and aerodynamic aspects.

The goal is now to integrate the aspect of uncertainty mentioned above into the optimization process. However, to do this, we must first identify the nature of these uncertainties, and decide how we should represent them. Several types of uncertainty have been identified and classified according to their nature.

Accounting for uncertainty has been studied in a number of research areas, but, until recently, in aeronautics research it was not possible to account for or quantify structural uncertainties within the optimization procedures due to the limitations of numerical tools and a lack of theoretical understanding of their impact within reliability studies. Engineers have therefore been forced to implement alternative procedures to simplify the integration of structural uncertainties into model development. The first studies on this topic in the aviation sector were only conducted in the 1990s, at which point this field of research began to produce tangible results.

In the case of optimization problems with probabilistic constraints, reliability-based optimization, which is extremely common in industrial contexts, replaces these probabilistic constraints with another deterministic optimization problem derived by techniques of approximation. The primary difficulty lies in evaluating the reliability of the structure, which is itself the result of another given optimization procedure. Reliability analysis is performed at the optimal point in order to determine the reliability index of the limiting state that is being considered.

This book presents the different aspects of fluid–structure interaction: vibroacoustics and aerodynamics, and the various numerical methods used to simulate them numerically.

One chapter is devoted to the question of model reduction in fluid–structure interaction problems. We begin by presenting dynamic substructuring methods in linear and nonlinear cases. We then give a description of the method of proper orthogonal decomposition for fluid flows. Finally, we present a modal synthesis method for solving large-scale coupled fluid-structure problems. This method couples a dynamic substructuring method of the type proposed by Craig and Bampton with an acoustic subdomain method based on an acoustic formulation of the velocity potential.

To account for uncertainty, one chapter presents concepts associated with reliability and its objectives and benefits in mechanics, methods for calculating the probability of failure, simulation methods such as the Monte Carlo and response surface methods, and approximate methods for analyzing the reliability and calculating the reliability index by the first-order reliability method (FORM) and the second-order reliability method (SORM). We then give a detailed presentation of the implementation of the latter approach in the context of a certain set of reliability-based optimization problems encountered when designing structures that interact with flowing fluids, with the goal of detecting the critical frequency bands that might cause the structure to experience damage or destruction.

Acknowledgments

We would like to thank everyone who has contributed in some way to this book, and in particular the engineering and PhD students at INSA in Rouen whom we have supervised over the past few years.

Abdelkhalak EL HAMI
Bouchaïb RADI
December 2016

1

Fluid–Structure Interaction

1.1. Introduction

Recently, several new problems have been formulated in the area of fluid–structure coupling, for example in the automotive industry with the dynamics of airbag inflation and fluid sloshing inside tanks; in aeronautics with the fluttering phenomenon affecting airplane wings, which involves a coupling between the vibrational dynamics of a structure and the flow of a fluid; and in the transportation industry with studies on noise reduction inside vehicles based on vibroacoustic analysis.

Each and every structure in contact with a fluid is subject to phenomena involving mechanical fluid–structure couplings to some extent. This kind of multiphysics coupling often significantly affects the dynamic behavior of mechanical systems. Taking it into account is one of the major challenges in calculating the dimensions of structures, especially when the objective is to ensure that their design meets the necessary safety requirements.

In this chapter, we will examine problems relating to the interaction of a structure with fluids both at rest and in flow. We will give a description of the motion of the fluid based on vibration theory, considering small vibrations in the structure and fluctuations in the pressure of the fluid around a stable equilibrium state, and we will present the relevant equations in the case of flowing fluids and the corresponding numerical methods for calculating couplings with dynamic structures.

1.2. Fluid–structure interaction problem

The mechanical coupling between the two media acts in both directions at their surface of contact: deformations in the structure resulting from the forces applied by the fluid flow modify the state of the fluid–structure interface; this affects the flow conditions of the fluid, which induces a change in the forces exerted on the structure at the interface, thus bringing the interaction cycle to a close.

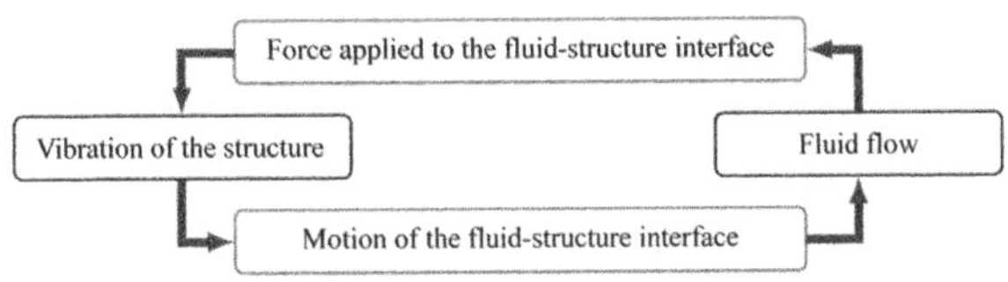

Figure 1.1. *Fluid–structure coupling mechanism. For a color version of this figure, see www.iste.co.uk/elhami/interactions.zip*

The fluid–structure interaction is described as the exchange of mechanical energy between a fluid and structure. This definition encompasses a wide range of problems. We can classify these problems using two criteria according to the physics of the problem at hand. The first criterion, proposed by Axisa [AXI 01], is based on the nature of the fluid flow. If the flow is negligible or non-existent, we say that the fluid is stagnant. Otherwise, we say that the fluid is flowing. In the first case, the objective is to describe small movements of the fluid and the structure around an equilibrium rest state. In these conditions, we choose to describe the dynamics of the interaction as a function of frequency; the equations describing the behavior of the structure and the fluid are written in terms of the reference (rest) state and generally lead to linear problems. In the second case, the objective is to establish a description of larger scale motion in the fluid and/or the structure. In these conditions, we choose to describe the dynamics of the interaction as a function of time; the equations describing the behavior of the structure and the fluid are written in terms of the current state of the system and generally lead to nonlinear problems.

The second criterion considers the coupling strength, which may be defined as the magnitude of the interactions or exchanges between the two media.

A coupling is said to be strong if there are high levels of exchange between the two media, i.e. the fluid has a significant impact on the structure, and vice versa. A coupling is said to be weak if the effect of one of the media dominates that of the other (Figure 1.2).

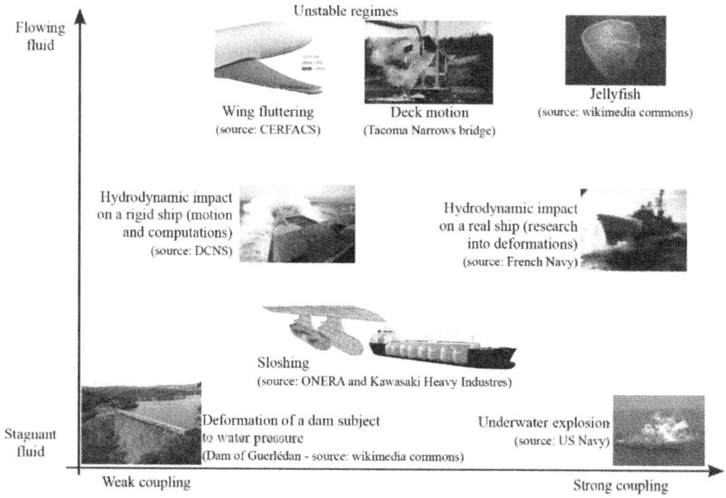

Figure 1.2. *Examples of fluid–structure interaction problems [GAU 11]. For a color version of this figure, see www.iste.co.uk/elhami/interactions.zip*

Three dimensionless numbers have been suggested to classify these problems [DEL 01]:

– The *mass number* M_A is defined as the ratio between the density of the fluid ρ_f and that of the structure ρ_s:

$$M_A = \frac{\rho_f}{\rho_s} \qquad [1.1]$$

This describes the significance of the inertial effects of the fluid and the structure. If its value is close to one, the inertial effects of the fluid are comparable to those of the structure, and so must be taken into account.

– The *Cauchy number* C_y is the ratio between the dynamic pressure and the elasticity of the structure, which is quantified by Young's modulus E.

$$C_y = \frac{\rho_f V^2}{E} \qquad [1.2]$$

This indicates the significance of the deformations induced by the flow. If this number is small, i.e. if the structure is rigid or the fluid velocity is small, structural deformations are negligible.

– The reduced velocity V_r is the ratio between the characteristic flow velocity and the velocity of wave propagation inside the structure:

$$V_r = \frac{V}{\sqrt{\frac{E}{\rho_s}}} = \frac{V}{c_s} \qquad [1.3]$$

If this number is large, the fluid dominates the problem from the perspective of time, and the dynamics of the structure are not important. By contrast, the dynamics of the structure increasingly dominate as this number tends to zero. If the number is close to 1, both dynamics carry similar weight in the problem.

These numbers are highly convenient for checking the importance of each phenomenon within the context of a given problem. However, as is the case for most dimensionless numbers, it is still difficult to define *a priori* threshold values applicable to all problems. In each problem, the large or small terms in the above will correspond to very different numerical values.

Using numerical simulations allows us to understand and predict the dynamic behavior of structures coupled with fluids, which is valuable in a number of industrial sectors. The numerical methods that we will use require us to solve the mathematical equations that model the behavior of the coupled fluid–structure system.

In general, the formulation of a coupled problem is based on the following description:

– the structure problem is formulated in terms of the displacement; the goal is to describe the behavior of the structure as a function of the displacement u, strain $\varepsilon(u)$, stress $\sigma(u)$, and to solve the equations of this dynamic to find the u, $\varepsilon(u)$ and $\sigma(u)$ fields in the structure domain;

– the fluid problem is formulated in terms of the pressure/velocity; the goal is to describe the behavior of the fluid as a function of the pressure p and the velocity v, to solve the equations of conservation of mass and momentum and to find the p and v fields in the fluid domain;

– at the fluid–structure interface, the mechanical exchanges are represented, on the one hand, by considering the force φ exerted by the fluid as a boundary condition for the structure problem and, on the other and, by considering the velocity $\frac{\partial u}{\partial t}$ imposed by the structure as a boundary condition for the fluid problem.

The energy exchanges between the fluid and the structure occur simultaneously. This needs to be taken into account by the numerical simulation. Coupled simulations can implement a single computational program to simultaneously solve the equations of the fluid and structure problems or alternatively can have two separate programs, one dedicated to the fluid problem and the other to the structure problem. The degree of complexity of the numerical simulation depends on the problem and methods of spatial and temporal discretization used to solve the equations of the problem.

Figure 1.3 proposes an overview of the most suitable general methods for simulating fluid–structure interaction problems:

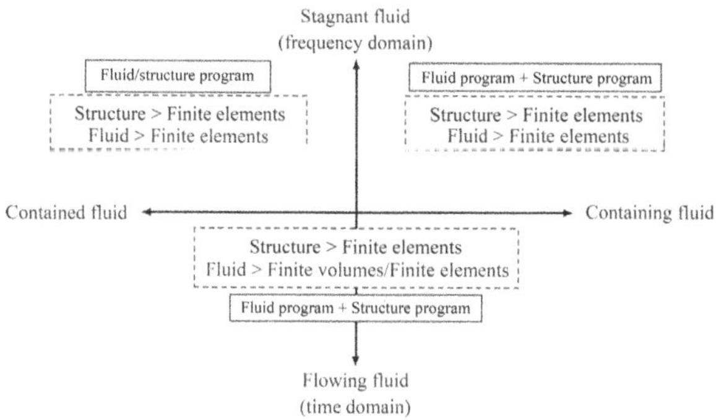

Figure 1.3. *General methods for numerically simulating fluid–structure interactions*

1.2.1. *Fluid–structure coupling methods*

There are several suitable coupling methods for the kinds of problems that we typically encounter. The following methods are used for stagnant fluids:

– The *decoupled method* finds the load or hydrostatic pressure on the structure, and then uses the results as an input to solve the deformation in the structure problem.

– *Acoustic fluid formulations (in terms of frequency)* allow small displacements around the equilibrium position of a structure to be determined. If the fluid is heavy, the vibrations of the structure and the fluid are strongly coupled. This coupling is reflected in the distinct natural frequencies of the modes, and the shapes of these modes. These methods use formulations that can be either non-symmetric (u, p) or symmetric (u, p, φ). They were proposed by Morand and Ohayon [MOR 95] and illustrated by Sigrist [SIG 11].

Frequency-based formulations of the fluid potential are applicable to problems with stagnant fluids, but can also be used to describe the elevation of a free surface subject to sloshing. The goal is to determine the motion of the free surface in order to find the pressure variations along the walls. These methods are similar to acoustic fluid formulations, which use either symmetric or non-symmetric expressions for the coupling equations, written as (u, p_0) and (u, h, φ) [SIG 11].

In this book, we will consider fluids that are flowing. It is important to note that the solutions of flow problems are based on Eulerian formulations. This kind of formulation is particularly well suited to the study of flows and greatly simplifies the process of solving the equations of the fluid problem.

Solving the deformation of a structure more naturally leads to a Lagrangian description. Expressing the interaction problem between a flowing fluid and a structure introduces an additional complication into the choice of formulation for the problem, as this formulation must be compatible with the models of both the fluid and the structure. Existing methods tackle this issue in different ways, which allows different levels of interaction to be taken into account.

– *Monolithic approaches* solve the stated problem as a single block. We can make a distinction between monolithic formulation and monolithic solving. The former category of approach describes and solves the problem with a formulation that is either Lagrangian or Eulerian. This gives the closest solution to the actual physical problem, but this solution is also the most difficult to formulate and solve. We can, for example, refer to the work by Morinishi and Fukui [MOR 12], who used an Eulerian formulation, and the work by Dermidzic and Muzaferija [DER 95], who additionally consider thermal effects. The latter type of approach, monolithic solving, describes the problem using both Eulerian and Lagrangian formulations simultaneously. It then merges and solves these two formulations in a single system. The methods commonly used for this are based on virtual domains with Lagrange multipliers or penalty-based methods [DIN 07, AQU 04], and immersed boundary methods [ZHA 07]. Hübner *et al.* [HÜB 04] implemented monolithic solving by computing the advancement of time with a finite element method that allows the fluid and structure equations to be solved simultaneously.

Both approaches are particularly suitable for strongly coupled problems. However, they are complex to implement, and currently they have been neither implemented nor validated by any widely available programs.

– *Partitioned approaches* separate the problem into fluid and structure subproblems, each of which is solved using a dedicated program. These two programs communicate with each other to exchange parameter values (pressure on the structure, displacement of the fluid–structure interface) required to perform the calculations. To improve the accuracy of the calculation, internal loops are iteratively executed at each time step to allow the parameter values exchanged by the fluid and structure solvers to converge. Using two distinct programs allows us to choose the most suitable solving method for each of the fluid and structure problems, and enables us to fully utilize all of the tools available to us. The difficulty with these methods lies in ensuring the quality of the coupling between the solvers and guaranteeing that all solvers converge properly.

– *Chained approaches* are a simplified form of partitioned methods. The same methods are generally used, but without internal iterations at each time step. These methods are suitable for weak couplings in which the deformations experienced by the structure are limited.

– *Rigid body methods* are used in cases where the transfer of mechanical energy causes the structure to move, but with negligible deformation. Since the motion is driven either by the flow or by gravity, there exists a coupling between the motion of the solid and that of the fluid. These types of problem are solved by finding the forces exerted by the fluid on the structure and using them to solve an equation of motion with six degrees of freedom (DOF; three translations and three rotations). There are two major methods for doing this. The first, known as 6 DOF, is commonly available in general purpose Navier–Stokes solvers. The second is the method of immersed boundaries, which is more complex and less widely employed.

The quality of the coupling depends not only on the quality of the solution of each system (fluid and structure), but also on the quality of the coupling algorithm [PIP 95]. Consequently, to achieve higher order couplings, the fluid and structure systems must have order greater than or equal to that of the coupling algorithm in time and space, which shows the importance of the choice of algorithm. These computation times are a consequence of the complexity of the system of fluid mechanics equations that must be solved. This system requires smaller time steps and a finer mesh than is usually necessary for the fluid component, which determines the number of unknowns of the problem. The computation time required by the dynamic mesh depends strongly on the amplitude of the motion of the structure and the algorithms used to calculate the displacement of the mesh nodes and recalculate the mesh if necessary.

To implement a partitioned coupling, we need coupling algorithms to connect the fluid and structure programs together, as well as the actual solvers themselves. These algorithms must allow the solvers to synchronize and exchange data. Synchronization consists of sending the right data at the right times, and executing iterative loops or establishing useful predictions. The data exchange protocol needs to guarantee that the information sent from one solver to the other is interpolated in an acceptable manner between the meshes of the two solvers. These two families of algorithms are further subdivided into the categories of temporal and spatial coupling.

1.2.2. *Temporal coupling*

Partitioned algorithms have the advantage of requiring less computation time than other methods of temporal coupling. However, they introduce an error into the information exchange between the fluid and structure components. This error is reduced when using models with predictors [FAR 98, PIP 95] and can be almost completely eliminated by using implicit algorithms.

Explicit synchronous method

The coupling algorithm is illustrated schematically in Figure 1.4. In this algorithm, the fluid and structure components are solved at the same time step using the prediction of the displacement of the mesh at time t^{n+1} as a function of the variables calculated at times t^n and t^{n-1} [SOU 10]. There are several possible forms of predictor; the choice of parameters affects the precision and stability of the models [SIG 10].

The forces calculated by the fluid program are not directly transmitted to the structure program; first, they are averaged: this allows the errors arising from the prediction of the displacement to be reduced. The averaging methods used for this are presented in [BEN 07].

Explicit asynchronous or shifted method

In this algorithm, the fluid iterations are shifted relative to those of the structure. The prediction is established for an intermediate time using a method similar to the one described above. The algorithm is presented in Figure 1.5.

Implicit method

This algorithm introduces an iterative loop to ensure that the predicted and calculated displacements converge (see Figure 1.6). This greatly reduces the errors arising from the prediction. Convergence criteria with higher levels of precision require longer computation times. This solver is more stable than explicit schemes.

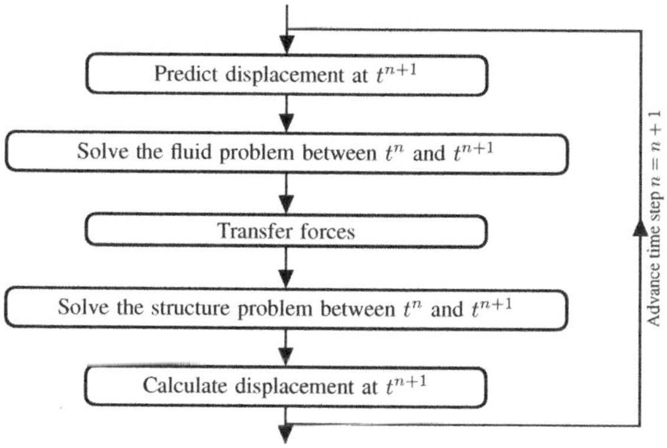

Figure 1.4. *Diagram of the explicit synchronous temporal coupling algorithm*

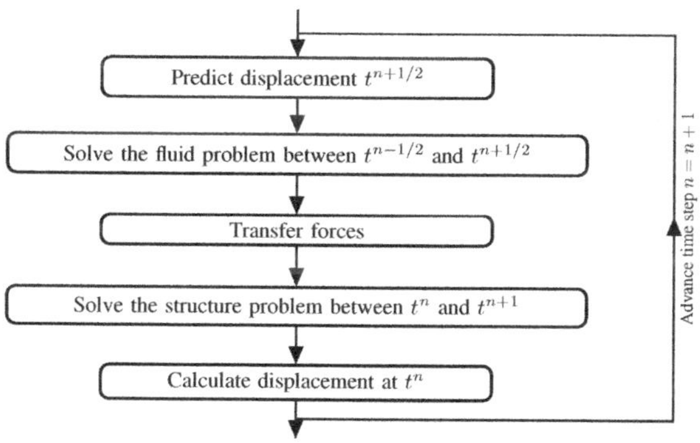

Figure 1.5. *Diagram of the explicit temporal asynchronous or shifted coupling algorithm*

Instability arises when the effects of added mass are non-negligible and the dimensions of the fluid domain are too large relative to the structure [CAU 04]. Hybrid methods have been developed to improve the stability of explicit methods [FER 05, FER 06, GER 03]. These methods determine the coupling forces implicitly and other coupling terms explicitly.

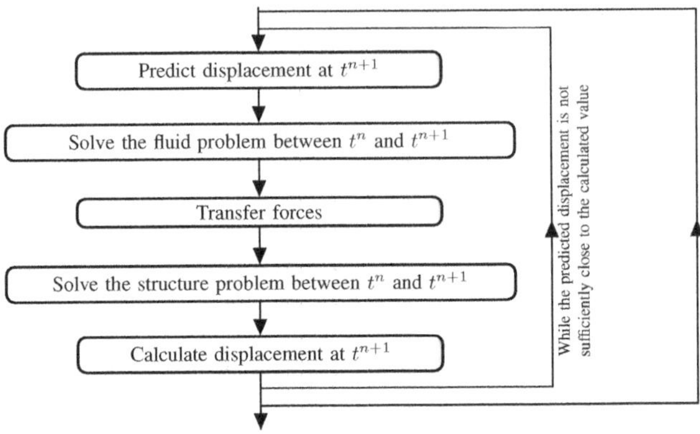

Figure 1.6. *Diagram of the implicit temporal coupling algorithm*

To evaluate the accuracy of the results, a comparative study of different models was performed on a system with two masses connected by a spring [BEN 07]. It was shown that the monolithic approach gives the best results, but that the results of this approach are similar to those obtained from the implicit method and same order as those obtained by explicit methods when the system includes damping. It was also shown that the correction step is not necessarily required for the forces in synchronous implicit and explicit methods, although it remains useful for asynchronous explicit methods. Piperno [PIP 97] showed that schemes with predictors are superior to schemes without predictors. On the other hand, an explicit scheme with a predictor does not perfectly conserve continuity in the displacements between the fluid and the structure.

Other studies [FAR 00] have shown that serial algorithms with predictors that are chosen to conserve continuity conditions for the velocity at the interface produce better results than other methods both with and without predictors. They also show that parallel models are less stable than the equivalent serial models, which further demonstrates that the choice of predictor significantly affects the quality of the coupling. Piperno and Farhat [PIP 00] show the benefit of predictors in both synchronous and asynchronous explicit methods, allowing higher order coupling algorithms that better conserve the energy balance of the coupled system to be found. Piperno *et al.* [PIP 05] and Piperno and Farhat [PIP 01] present a comparison criterion for temporal coupling models based on the energy exchanged at the interface.

1.2.3. *Spatial coupling*

The forces exerted by the fluid on the structure and the displacements induced by the structure on the fluid are exchanged through meshes of nodes. The difficulty with spatial coupling is that information must be transmitted losslessly between these meshes without introducing an error [FAR 98, MAM 95]. In most cases, the structure and fluid meshes do not coincide. Spatial coupling can therefore be thought of as two separate phases: first, the nodes of one mesh are projected onto the other, and then the parameter values are determined at these nodes.

Projecting nodes

The simplest method is to project the nodes of one mesh onto the other along the normal of one of the two meshes. The choice of normal and projected mesh is significant, as it will partially determine the quality of the spatial coupling [MAM 95].

Figure 1.7 shows an example in which the nodes of a structure mesh have been projected onto a fluid mesh. Note that projection is not possible for the structure node S_5. The value at this node is reassigned to the closest fluid node, which is F_8. There are two possible projections, d_1 and d_2, for the structure node S_2. We choose the smaller of the two, which in this case is d_2. The fluid element bounded by the nodes F_2 and F_3 does not contain any information originating directly from the structure.

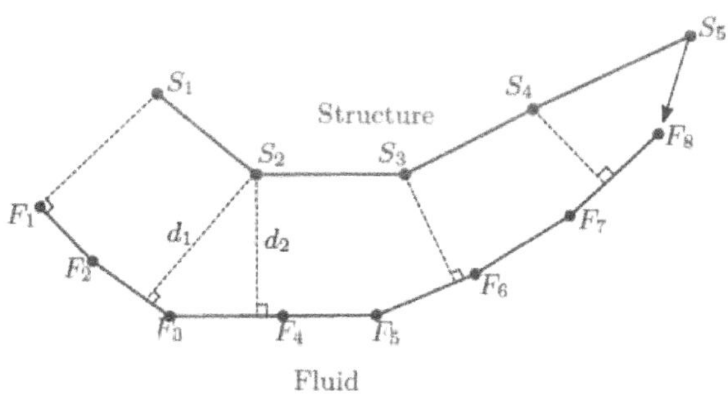

Figure 1.7. *Example of projection between two meshes. For a color version of this figure, see www.iste.co.uk/elhami/interactions.zip*

In Figure 1.7, we chose to project the structure nodes onto the fluid mesh. This solution results in fewer projections, since the structure mesh is usually less dense; however, the reverse solution is also possible. Projecting the nodes of both meshes onto an intermediate coupling element has also been suggested [FAR 98].

To improve the implementation of this method, we can subdivide the interface into areas known as *buckets*. Each area may then be processed independently of the others, with special algorithms to handle cases where only one of the two interfaces is present in a given area. Figure 1.8 shows the concept of buckets for the previous example.

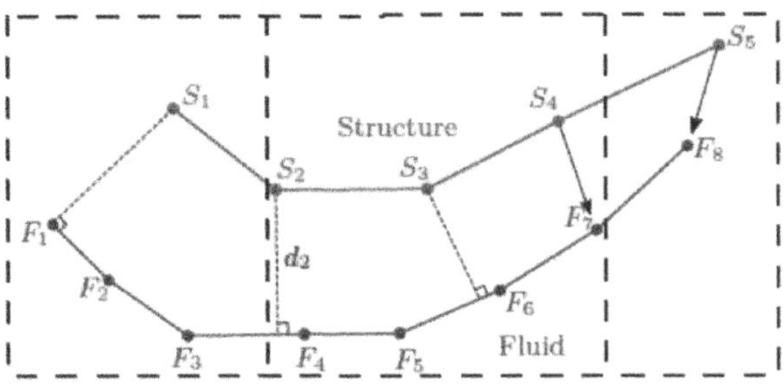

Figure 1.8. *Example of a projection with a buckets method. For a color version of this figure, see www.iste.co.uk/elhami/interactions.zip*

Determining the parameter values at each node

This presents the methods used by Ansys MFX to determine the parameter values at each node. The first method, conservation of profile, linearly interpolates parameter values at the projected nodes using an idea similar to that proposed by Piperno [PIP 97]. This method conserves the integral of the exchanged parameter value I (see Figure 1.9). The second method, global conservation, determines the parameter values at the projected nodes using a shape function. It conserves the sum of the parameter values taken over the interface (see Figure 1.10).

As shown by Figures 1.9 and 1.10, the method of determining the parameter values at the nodes strongly affects the error introduced into the calculation by spatial discretization. The global method is usually better for discrete values that do not need to be conserved exactly. However, for forces and fluxes, keeping the integrals equal is important to ensure that the energy exchanged by the fluid and the structure is conserved. Therefore, in these cases, the method of conservation of profile is better.

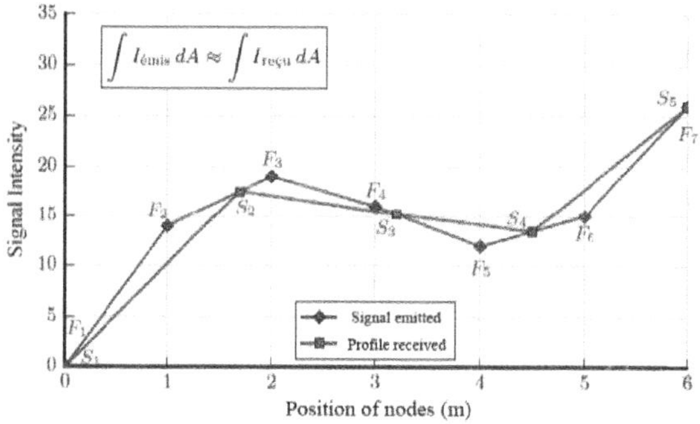

Figure 1.9. *Method of profile's conservation. For a color version of this figure, see www.iste.co.uk/elhami/interactions.zip*

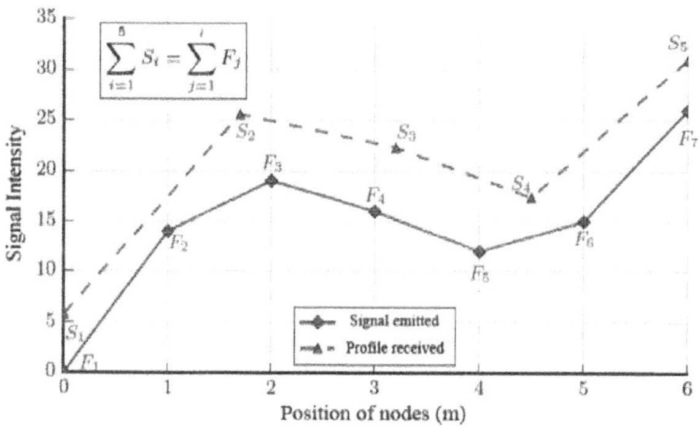

Figure 1.10. *Method of global conservation. For a color version of this figure, see www.iste.co.uk/elhami/interactions.zip*

General grid interface method

One variant of the global method is the *general grid interface* (GGI) method. This method incorporates the steps of projection and interpolation. Figure 1.11 shows how this method works in a simple example.

First, the faces of each element of the interface are divided by the number of nodes on the face. These faces, which are described as IP, are converted into polygons composed of rows and columns of pixels with a resolution of 100×100. The polygons thus created on the emitting face (which in this case is the fluid mesh) are intersected with those on the receiving face. This constructs the control surfaces. These control surfaces may then be used to exchange parameter values.

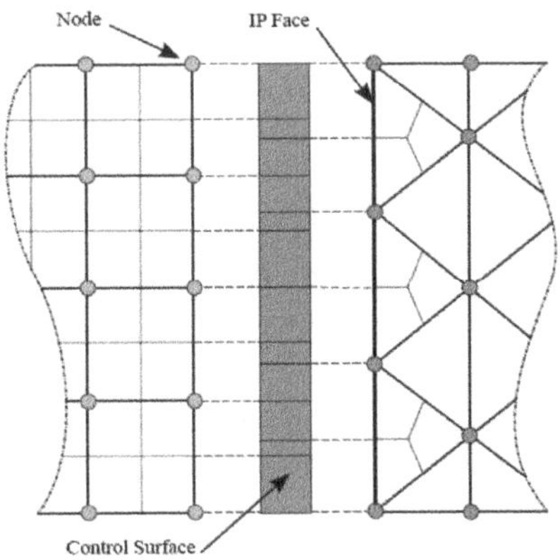

Figure 1.11. *General grid interface (GGI) exchange method. For a color version of this figure, see www.iste.co.uk/elhami/interactions.zip*

Other methods are possible, such as the closest neighbor method, which assigns the value of the projected node to the closest node in the receiving mesh. Other interpolation methods can also be used. Using a consistent method (i.e. the truncation error tends to zero as the grid spacing decreases) significantly improves the robustness and accuracy of the results [FAR 98].

1.3. Vibroacoustics

In many situations, sound is created by the interaction between the vibration of a structure and its acoustic radiation. This field of study, known by the name of vibroacoustics, has an extremely wide and diverse range of applications, from musical instruments to ultrasound scanning and active vibration control to the transmission of vibrations and sound between media with distinct propagation properties, or in the form of acoustic radiation. Studying the history of a field allows us to understand how both teaching and research have progressed within it. The classical approach to teaching vibrations focused on the mechanical response of a structure subjected to an idealized stimulus, neglecting to consider the surrounding environment. Later, the properties of fluids satisfying the assumption of incompressibility (sound waves do not propagate) were taken into account. Finally, the study of coupling phenomena and fluid–structure interactions led to vibroacoustics as we know it today.

Several different reasons motivated these shifts in perspective:

– difficulties encountered in the fields of nuclear, spatial and aeronautical research, and later also in land-based forms of transport, inspired a large amount of research into improving the reliability of mechanical systems;

– the development of increasingly powerful software tools in the 1970s made it possible to apply a number of very old conceptual tools (ranging from Helmholtz's radiation integral from 1865 to Courant finite elements from 1940). These tools also enabled the development of increasingly refined models and led to highly efficient experimental methods for signal processing;

– the introduction of a number of national and international regulations created the need to reduce noise levels in both the industrial and construction sectors;

– competition prompted manufacturers to research, innovate and adopt new technologies, even in the absence of regulations (for example in the area of internal noise levels in cars, trains and planes).

The concepts of acoustic quality and sound design currently represent the culmination of these developments. By the very nature of the couplings on which it is based, the field of vibroacoustics requires a cross-disciplinary approach. Whether to study a musical instrument, or to estimate the level of internal noise in an airplane, the researcher must take into account each of the various sources and how they correlate with each other, the propagation pathways of the vibrational and acoustic energy, the properties of materials and systems assumed to be decoupled, and the type of coupling involved in order to ultimately formulate a description of the vibrational response of these structures, the pressure, the acoustic intensity, and the radiated power. When modeling the propagation through real environments, we can distinguish between two different cases. The first is the case of an interface between two propagating media. The reader would therefore be correct in suspecting that the second is that of an interface between one propagating and one non-propagating medium. In fact, the distinction is slightly more subtle, as in the first case we are only interested in propagating media of same type (two fluids), whereas in most cases physical problems involve propagation between two propagating media of different types, typically a fluid and an elastic solid. The complexity of vibrational coupling phenomena involving waves with different types of behavior makes them highly difficult to understand.

1.3.1. *Vibrations of three-dimensional solids*

Wave propagation in elastic media differs significantly from wave propagation in fluid media. For example, in a fluid, there is only one type of wave, whereas in an infinite homogeneous and isotropic elastic solid there can be both longitudinal waves, in which the particles move in the direction of propagation, and independent transverse waves, in which the particles move orthogonally to the direction of propagation. These two types of wave are coupled through phenomena that occur near discontinuities (geometrical or mechanical). Most internal source mechanisms generate both types of wave simultaneously.

Mechanical wave propagation equation

Consider a homogeneous and isotropic solid with density ρ, Young's modulus (E), Poisson's ratio (ν) and Lamé coefficients (λ, μ), satisfying:

$$\lambda = \frac{E\nu}{(1-2\nu)(1+\nu)} \;; \quad \mu = \frac{E}{2(1+\nu)}$$

Similarly to the fluid case, we are interested in how a perturbation propagates throughout the solid. In most cases, this hypothesis enables us to neglect all but the linear part of the motion, and we can assume that the displacements and strains that occur within the solid are sufficiently small. Given these assumptions, we know that the relation between the strain experienced by each point in the solid and the stress applied to these points is given by Hooke's law, which in the homogeneous and isotropic case may be written as:

$$\sigma = \lambda \mathrm{Tr}(D) + 2\mu D \qquad [1.4]$$

Recall the equation of conservation of momentum (with Einstein's summation convention for repeated indices) in terms of Euler variables:

$$\rho\frac{\partial^2 u_i}{\partial t^2} + \frac{\partial(\rho u_i u_j - \sigma_{ij})}{\partial x_j} = F_i \qquad [1.5]$$

where u_i are the components of the displacement field of the solid, σ_{ij} are the components of the stress tensor and F_i are the components of the applied force.

Assuming that the strain is small, we can linearize this equation, which amounts to neglecting the $u_i u_j$ term. From Hooke's law, we deduce that:

$$\begin{aligned}\frac{\partial\sigma_{ij}}{\partial x_j} &= \frac{\partial\lambda d_{ll}\delta_{ij} + 2\mu d_{ij}}{\partial x_j}\\ &= \lambda\frac{\partial d_{ll}}{\partial x_i} + 2\mu\frac{\partial d_{ij}}{\partial x_j}\end{aligned}$$

where

$$d_{ij} = \frac{1}{2}\left(\frac{\partial u_i}{\partial x_j} + \frac{\partial u_j}{\partial x_i}\right) \qquad [1.6]$$

Let:

$$\begin{aligned}\frac{\partial \sigma_{ij}}{\partial x_j} &= \lambda \frac{\partial^2 u_j}{\partial x_i \partial x_j} + \mu\left(\frac{\partial^2 u_j}{\partial x_j^2} + \frac{\partial^2 u_j}{\partial x_i \partial x_j}\right)\\ &= (\lambda+\mu)\frac{\partial^2 u_j}{\partial x_i \partial x_j} + \mu \frac{\partial^2 u_2}{\partial x_j^2}\end{aligned}$$

We substitute this expression into the law of conservation of momentum. We obtain

$$\rho \frac{\partial^2 u_i}{\partial t^2} - (\lambda+\mu)\frac{\partial^2 u_j}{\partial x_i \partial x_j} - \mu \frac{\partial^2 u_2}{\partial x_j^2} = F_i \qquad [1.7]$$

This is the linearized equation of elastic wave propagation.

1.3.2. *Acoustics of fluids*

In order to ensure that the mathematical problem that models the propagation of sound is well posed when the domain in which propagation occurs contains obstacles, we need to introduce boundary conditions at the edges in the wave equation. If the domain is infinite, we impose "outgoing wave" type boundary conditions to satisfy the principle of conservation of energy. This essentially says that energy cannot come from infinity. In particular, when we calculate the Green's functions, we will choose the elementary solution that satisfies this condition.

The vibrational behavior of a fluid (small fluctuations around an equilibrium state) assumed to be homogeneous, perfect and initially at rest may be described by a scalar variable that characterizes the state of the fluid. The problem equations are obtained by linearizing the general equations for the fluid flow [AXI 01, GIB 86, MOR 95] (Navier–Stokes equations). These equations express the principles of conservation of mass and momentum, and may be written in the following forms:

– conservation of mass

$$\frac{\partial \rho}{\partial t} + \frac{\partial(\rho v_j)}{\partial x_j} = 0 \qquad [1.8]$$

– conservation of momentum

$$\frac{\partial(\rho v_i)}{\partial t} + \frac{\partial(\rho v_i v_j)}{\partial x_j} = -\frac{\partial p}{\partial x_i} + \mu \frac{\partial^2 v_j}{\partial x_j^2} \qquad [1.9]$$

where p and $v = (v_i)$ are the pressure and velocity fields of the fluid.

These equations are supplemented by the equation of state of the fluid, which may, for example, be written as:

$$p = \psi(\rho)$$

Linear vibrations of the fluid may be described using formulations based on the displacement [BER 04, HAM 78], the velocity potential [SIG 11], the displacement potential [BOU 87, SAN 88] or the pressure [MAR 78, SIG 07]. In the last of these cases, we examine changes in the pressure p and density ρ of a non-viscous fluid (i.e. such that $\mu = 0$) starting from some initial state. These values can be written using the following decomposition:

$$p(x,t) = p_0(x) + p'(x,t) \qquad \rho(x,t) = \rho_0(x) + \rho'(x,t)$$

in which we separate the fluctuating component (p', ρ') from the steady component (p_0, ρ_0), which characterizes the initial state of the fluid.

1.3.3. *Numerical methods for calculating a structure coupled with a stagnant fluid*

Modeling assumptions

In this section, we will discuss the numerical methods that may be used to solve a coupled fluid/structure problem in the case where the fluid is stagnant. The analysis framework is based on the vibrations in the structure and the fluid; in the fluid, we consider the propagation of waves arising from gravitational effects (at low frequencies) or compressibility (at high frequencies).

The state of the coupled systems is described by the value of the displacement field in the structure domain and by the value of the pressure field in the fluid domain. The method of finite elements is useful for describing vibrations in elastic structures; it is equally suitable for describing vibrations in a fluid (gravitational or compressional waves) for continuous fluids (bounded domains) contained inside a structure; in this case, finite element (structure)/finite element (fluid) coupling can be used to solve the coupled problem. This type of method can be implemented using a finite elements approach: a coupling operator gives a rigorous description of the interaction in terms of action/reaction.

In the case of a fluid that contains a structure (non-bounded domains), the method of finite elements ceases to be applicable: the boundary element method may instead be used to take into account effects arising from the fact that the fluid domain is infinite. Finite element (structure)/boundary element (fluid) coupling can be used to solve the coupled problem: this category of method generally requires two distinct computational programs to be coupled together (one program for the finite elements and another for the boundary elements).

Vibrational analysis of elastic structures

In linear problems involving small movements, the vibrations of the structure are expressed in terms of frequency; the local equilibrium equation, which is obtained from the fundamental principle of dynamics without external forces, can be written as follows in a Cartesian coordinate system:

$$\omega^2 \rho_S u_i + \frac{\partial \sigma_{ij}(u)}{\partial x_j} = 0 \quad \text{in} \quad \Omega_S \qquad [1.10]$$

Initially, the structure is considered on its own, i.e. without the fluid, so that $\Gamma = \emptyset$. The boundary conditions on the constrained boundary and the stress-free boundary may be written as:

$$u_i = 0 \quad \text{on} \quad \Gamma_{S_0} \qquad [1.11]$$

$$\sigma_{ij}(u) n_j^S = 0 \quad \text{on} \quad \Gamma_{S_\sigma} \qquad [1.12]$$

Grouping together equations [3.155]–[3.157], we deduce an equation for the displacement field only (Navier equation):

$$\rho_S \omega^2 u_i + (\lambda + \mu) \frac{\partial}{\partial x} \left(\frac{\partial u_j}{\partial x_j} \right) + \mu \frac{\partial^2 u_i}{\partial x_j \partial x_j} = 0 \qquad [1.13]$$

In the general case, it is not possible to find an analytical solution to equation [3.160] with boundary conditions [3.158] and [3.159]; an approximate solution can be obtained by discretizing a formulation based on weighted integrals that is equivalent to the initial problem. This formulation is derived using the method of test functions by considering an arbitrary displacement field δu, known as the virtual displacement field, that satisfies the boundary condition [3.158]. Multiplying [3.157] by δu and integrating over the domain, we can write:

$$-\omega^2 \int_\Omega \rho u_i \delta u_i d\Omega - \int_\Omega \frac{\partial \sigma_{ij}(u)}{\partial x_j} \delta u_i d\Omega = 0 \qquad [1.14]$$

Using generalized integration by parts and taking into account the boundary conditions [3.158] and [3.159], we deduce the following integral formulation satisfied by any field δu:

$$-\omega^2 \int_{\Omega_S} \rho_S u_i \delta u_i d\Omega_S + \int_{\Omega_S} \sigma_{ij}(u) \varepsilon_{ij}(\delta u) d\Omega_S = 0 \qquad [1.15]$$

This integral formulation can then be discretized using numerical methods. While many approaches are possible (finite differences, boundary elements, distinct elements, etc.), the finite element method [BAT 82, ZIE 89, BEL 00] is currently the most widely employed in computational programs.

Vibrational analysis of stagnant fluids

The Helmholtz equation describes the propagation of sound waves as a function of frequency:

$$-\frac{\omega^2}{c^2} p - \frac{\partial^2 p}{\partial x_i \partial x_i} = 0 \quad \text{in} \quad \Omega_F \qquad [1.16]$$

This equation comes with boundary conditions of the form:

$$\frac{\partial p}{\partial x_j} n_j^F = 0 \quad \text{on} \quad \Gamma_{F_\pi} \qquad [1.17]$$

$$p = 0 \quad \text{on} \quad \Gamma_{F_0} \qquad [1.18]$$

The boundary condition [1.17] models the presence of a fixed wall bounding the fluid domain: it says that the fluid flux at the wall is zero in the normal direction, but that fluid motion is possible in the tangential direction and independently from the motion of the wall (perfect fluid). This condition is also used to formulate a symmetry condition for the fluid problem. The boundary condition [1.18] models the existence of an acoustic free surface: it says that on this free surface, the (absolute) pressure is fixed at a given value by external factors, and so the pressure fluctuations are equal to zero. This condition is also used to formulate an antisymmetry condition for the fluid problem. Analytical solutions can sometimes be found for problems determined by the above equations, usually for elementary geometries. In the general case, approximate solutions can be found using numerical methods. The finite element method is particularly well suited to the problem as stated here; it requires us to pass via the weighted integral formulation, which is equivalent to [1.16]–[1.18]. This formulation is derived using the method of test functions by considering an arbitrary pressure field δp that satisfies the boundary condition [1.18]. By multiplying [1.16] by δp and integrating over the domain using generalized integration by parts and taking into account the boundary conditions [1.17] and [1.18], we deduce the following integral formulation satisfied by any field δp:

$$-\omega^2 \int_{\Omega_F} \frac{p\delta p}{c^2} d\Omega_F + \int_{\Omega_F} \frac{\partial p}{\partial x_i}\frac{\partial \delta p}{\partial x_i} d\Omega_F = 0 \qquad [1.19]$$

1.4. Aerodynamics

In general, aeroelasticity is the area of applied mechanics that deals with the motion of deformable bodies in gaseous flows (or hydroelasticity for fluid flows). Although aeroelasticity was originally studied in connection with aeronautics, because the types of problem considered by aeroelasticity proved to be critical in the early days of propulsion-based flight, aeroelastic phenomena have also played a very important role in other fields of applied science. For instance, civil engineering has a history of undertaking projects with ever bolder designs and ever higher flexibility (buildings, bridges, towers, power lines, etc.). Similarly, when designing turbomachines, hydroelectric projects, land or sea vehicles, etc., aeroelastic and hydroelastic problems are becoming ever more relevant. Hence, the field of aerolasticity remains highly significant in scientific and industrial research to this day.

Nevertheless, aerospace manufacturing is the area in which aeroelastic phenomena are most prominent. Consequently, since the Second World War, great emphasis has been placed on aeroelastic phenomena when building high-speed (transonic and supersonic) and large-scale aircraft (gliders, etc.). In particular, for high-speed aircraft, it has been necessary to modify the geometry of the wings and introduce servocontrols, which has inspired new problems. The emergence of new materials (such as composite materials) has also prompted an increase in scale (and flexibility) in all categories of aircraft, which increases the significance of aeroelastic effects.

Aeroelasticity belongs to the family of phenomena with interactions between a flow and a structure. One characteristic feature of this kind of phenomenon is that, in some cases, the fluid can supply an indefinite quantity of energy to the structure, which leads to system instability. There is another way to characterize these phenomena. We can distinguish phenomena in which:

– the fluid and the structure have comparable densities. This is, for example, the case with immersed structures and physiological flows;

– the fluid and the structure have very different densities: usually, the density of the fluid is lower than that of the structure.

The aeroelastic phenomena that we will consider here belong to the second category, as the fluid is significantly "lighter" than the structure. One final property can be used to characterize aeroelasticity problems: the distinct spatial separation between the fluid and the structure. It is important to emphasize this feature, as it plays an important role in the choice of numerical method when computing these phenomena. Aeroelasticity may therefore be defined as the study of the elastic behavior of structures whose motion within the flow generates induced stress. This topic combines three disciplines:

– aerodynamics, to predict the forces experienced by the structure;

– elasticity, to determine alterations to the structure (displacements and deformations);

– structural dynamics, to determine the inertia matrices and modal properties (modes, natural frequencies) and in some cases the inertial forces (for motion involving non-uniform acceleration).

We will study two major phenomena:

– static phenomena: the structure experiences strain as a result of the aerodynamic forces that it applies to itself.

– dynamic phenomena: the fluid supplies energy to the structure, which may either amplify oscillatory motion or cause the system to break up if the maximal tolerances are exceeded. This phenomenon is called fluttering.

1.4.1. *Aeroelastic problems*

Aeroelastic problems arise from the interaction of different types of force:

– elastic forces, which are structural in origin;

– inertial forces, which are also structural;

– aerodynamic forces, which are induced by strain (steady or oscillatory) in the structure, and are the result of external perturbations.

Aeroelastic problems only arise because the aircraft structure is not perfectly rigid when an air flow is applied to it. Aircraft structures are always flexible to a greater or lesser extent, and this flexibility is the underlying cause of each observable aeroelastic phenomenon.

Inertial effects can also play a very important role, and so we will differentiate between *dynamic aeroelasticity*, which involves all three of the forces listed above, and *static aeroelasticity*, which only involves elastic and aerodynamic forces.

Classically, aeroelastic phenomena are classified using Collar's triangle of forces (Figure 1.12). The three types of force arising from motion (elastic, aerodynamic and inertial) are represented by the three vertices of the triangle. Each aeroelastic phenomenon can be situated on this diagram according to how it relates to each vertex. For instance, phenomena relating to dynamic aeroelasticity are located at the center of the triangle, whereas effects relating to static aeroelasticity are located on the left-hand side. The right-hand side groups together phenomena that only involve aerodynamic and inertial forces, such as the dynamics of rigid aircraft studied in *flight mechanics*. The base of the triangle corresponds to vibrational problems from *structural dynamics*.

The most important phenomenon of static aeroelasticity is *divergence*: above a certain speed, the equilibrium between aerodynamic forces and elastic spring forces becomes unstable, which leads to a sudden collapse of the structure.

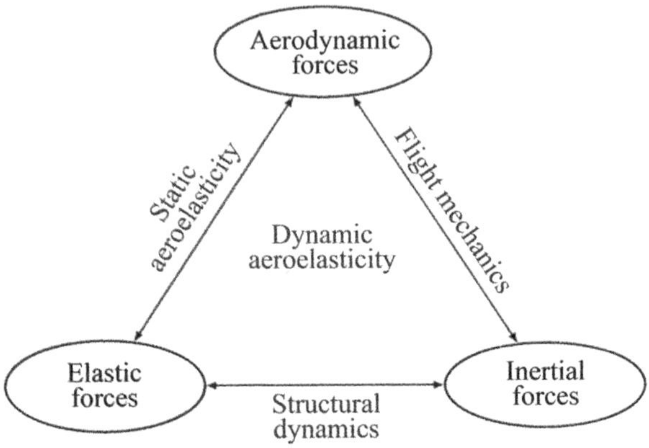

Figure 1.12. *Aeroelasticity triangle*

Dynamic aeroelasticity includes, in particular, the following phenomena:

– *fluttering*: this is an example of dynamic instability, coupling an unsteady flow with the vibratory motion of the structure, which either dampens or amplifies this vibration.

– *dynamic response*: the flexibility of the aircraft can significantly alter its response to atmospheric perturbations (gusts, turbulence) and fast maneuvers.

From a mathematical perspective, the study of instability is complementary to that of dynamic response. Indeed, instability conditions are generally determined by the presence of non-trivial solutions in a system of homogeneous equations, whereas the dynamic response is obtained by solving these same equations after adding a source term.

In air flows around a structure in motion, three (or more) dimensionless quantities can be used to classify the nature of the mechanisms that can be expected to occur:

– *The Reynolds number* is the ratio between the convected kinetic energy and the energy that is emitted and dissipated by friction in the fluid, defined as:

$$Re = \frac{VL}{\nu}$$

where L is the wing length (in m), V is the wind speed (in m/s) and ν is the kinematic viscosity (in m/s).

– *The Strouhal numberl* is related to the instability of boundary layers, and represents the ratio between the characteristic length and the distance travelled by a fluid particle at the characteristic speed over one characteristic period of the flow:

$$St = \frac{L}{V_{ref} T_{ref}}$$

– *The reduced frequency* f_r determines which mode of operation is most appropriate for studying fluid–structure interaction phenomena in the presence of a flow. It is defined as the ratio between two characteristic periods of the system, the first of which is the time taken by a fluid particle to travel across the structure (T_f), and the second of which is the natural period of vibration of the structure ($T_s = \frac{1}{f_s}$). We therefore have that:

$$f_r = \frac{T_f}{T_s} = \frac{\frac{L}{V}}{T_s} = \frac{L f_s}{V}$$

The inverse of the reduced frequency is known as the reduced speed. f_r is sometimes also defined in terms of the angular frequency instead of the natural frequency. From the perspective of aeroelastics, three different regimes can be distinguished according to their reduced frequency:

– if $f_r \gg 1$, physically this means that the motion of the fluid is slow relative to the motion of the structure. This domain is studied by the field of acoustics, or aeroacoustics if the fluid velocity is taken into consideration;

– if $f_r \simeq 1$, we are in the regime of strong interactions. In this regime, there exists some form of resonance, since each medium exerts stress upon the other with similar frequencies. This is the most complex regime and by far the most difficult to solve;

– when $f_r \ll 1$, the motion of the structure is slow relative to that of the fluid, which allows us to estimate the stresses applied to the structure by means of a permanent flow. This greatly simplifies the analysis and the formulations involved. This is the domain of quasi-steady aeroelasticity, which is extremely common in industrial applications, and for which a great number of reliable strategies have been developed and perfected.

When the reduced frequency takes very low values, we can justifiably assume that the oscillations do not produce any effect on the flow around the structure, and that changes in the motion of the solid are instantaneously applicable to the flow and the stresses exerted by it. Quasi-steady theory uses static aerodynamic coefficients, i.e. coefficients evaluated on a stationary structure, to calculate changes in the fluttering.

1.4.2. *Aerodynamic loads*

Aerodynamic loads can be evaluated locally on a structure using the pressure distribution at the wall, or alternatively using a stress torsor.

Action of the pressure and pressure coefficients

The pressure acts according to Cauchy's principle along the normal to the wall and is proportional to the static pressure p. The elementary force dF exerted by a surface element with unit normal vector N may simply be written as:

$$dF = -pNds$$

The minus sign is included by convention and depends on the choice of direction for the vector N.

However, the action of the static pressure is not the only force applied locally. Viscous fluid flow creates a boundary layer characterized by its velocity gradient at the wall. Given the assumption of a Newtonian fluid, this gradient is linear and so results in shear stress. This adds an additional friction term to the pressure force that is ultimately due to the viscosity of the fluid.

In practice, for non-profiled structures, this friction term is negligible relative to the action of the pressure. But for structures with specific profiles such as the wings of an aircraft, the friction term contributes in equal measure to the drag force, and cannot be neglected.

In practice, the pressure is written in dimensionless form. Since it depends on the flow velocity, it is often expressed as a pressure coefficient C_p given by:

$$C_p = \frac{p - p_{ref}}{q_{ref}}$$

where $q_{ref} = \frac{1}{2}\rho V_{ref}^2$.

The characteristic values are averages that are usually determined prior to modeling, for example using a Prandtl tube. The reference pressure that is chosen in practice is often the static pressure as determined in advance. Hence, the pressure coefficient will never have a value greater than 1, as it follows from Bernouilli's theorem for steady flows that the maximum possible value of p is the stagnation pressure p_0, which indeed has a maximum possible value of 1.

At the rear end of non-profiled structures, the pressure coefficient is negative, and does not have a theoretical lower bound. The drag of this type of structure is primarily caused by a low-pressure area in a phenomenon known as the base flow effect.

The pressure coefficient is a parameter that is generally independent of the flow rate, and depends solely on the geometry of the structure. It can depend on the Reynolds number, in particular due to variations in the positions of the points at which the boundary layer detaches and reattaches on non-profiled structures.

Aerodynamic forces and moments

If the pressure distribution along the wall is known, it is relatively straightforward to integrate it to find the global forces. But in some situations we might also wish to measure these forces directly using aerodynamic moments. In this case, it is essential to first define the reference point with respect to which the torque is defined. Researchers in aerodynamics traditionally express the forces in terms of wind coordinates (by convention the Eiffel coordinate system), but when computing the structure it is often more convenient to work in structure coordinates (known as the Lilienthal coordinate system) [DES 08]. Using the notation in Figure 1.13, the wind coordinate system with axes (O, D) and (O, L) is obtained by rotating through angle α from the structure coordinate system with axes (O, x) and (O, z). In the three-dimensional case, we can add the (O, y) axis, which is perpendicular to the plane shown in the figure. All of these are examples of right-handed coordinate systems. The angle α is the angle of incidence.

The three-dimensional stress torsor in structure coordinates is composed of three forces:

- F_x: drag force;
- F_y: cross-wind force;
- F_z: lift force;

and three moments.

- M_x: rolling moment;

– M_y: pitching moment;

– M_z: yawing moment.

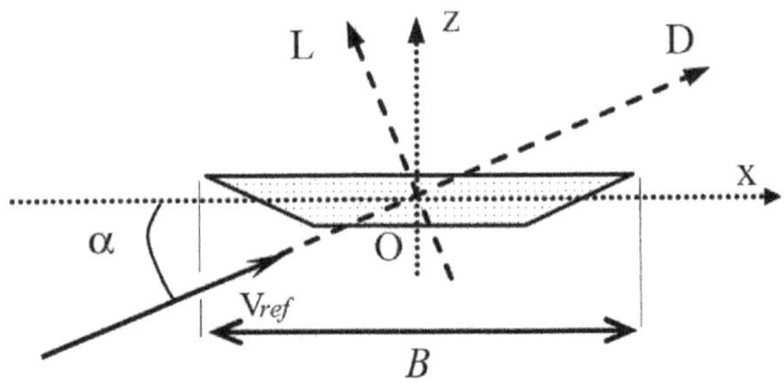

Figure 1.13. *Notation of two-dimensional coordinate systems*

In the two-dimensional case, if we consider the vertical plane, only the drag and lift forces and the pitching moment are significant. The aerodynamic forces are usually given in the form of dimensionless coefficients, which makes it easy to transpose forces measured on models over to physical structures. We thus define the force coefficients as:

$$C_x = \frac{F_x}{\frac{1}{2}\rho V_{ref}^2 S} \quad [1.20]$$

$$C_y = \frac{F_y}{\frac{1}{2}\rho V_{ref}^2 S} \quad [1.21]$$

$$C_z = \frac{F_z}{\frac{1}{2}\rho V_{ref}^2 S} \quad [1.22]$$

where S is a reference surface, for example the deck surface of a bridge.

When the forces are given in two dimensions, the forces are given per unit length and the reference surface S is usually replaced with the width B of the structure. The drag and lift coefficients in wind coordinates are written as C_D and C_L, respectively.

The moment coefficients are given by:

$$C_L = \frac{M_x}{\frac{1}{2}\rho V_{ref}^2 SB} \quad [1.23]$$

$$C_M = \frac{M_y}{\frac{1}{2}\rho V_{ref}^2 SB} \quad [1.24]$$

$$C_N = \frac{M_z}{\frac{1}{2}\rho V_{ref}^2 SB} \quad [1.25]$$

In two dimensions, only the pitching coefficient C_M is useful; it characterizes the torque per unit length. In general, we choose the axis that passes through the center of the structure to define the pitching coefficient.

1.4.3. *Problem equations*

To model aeroelastic phenomena, we have to simultaneously solve the fluid mechanics equations and the structural mechanics equations.

Flow equations

The description of fluid flows is generally based on an Eulerian formulation that allows us to examine the flow properties in each of the different regions of the fluid domain. We will restrict ourselves to the case of linear viscous fluids, commonly known as Newtonian fluids, which are by far the most important in practical applications. Newtonian fluids are characterized by their behavioral equations, such as the fact that the relation between the viscous stress σ and the strain is linear:

$$\sigma_{ij} = \mu\left(\frac{\partial v_i}{\partial x_j} + \frac{\partial v_j}{\partial x_i} - \frac{2}{3}\frac{\partial v_k}{\partial x_k}\delta_{ij}\right) - p\delta_{ij} \quad [1.26]$$

where v_i is the velocity vector with Cartesian coordinates x_i, p is the pressure, μ is the dynamic viscosity and δ_{ij} is the Kronecker symbol.

The laws of conservation of mass, motion and energy may therefore be written as:

$$\frac{\partial \rho}{\partial t} + \frac{\partial(\rho v_i)}{\partial x_i} = 0 \quad [1.27]$$

$$\frac{\partial(\rho v_i)}{\partial t} + \frac{\partial(\rho v_i v_j)}{\partial x_j} = \frac{\partial}{\partial x_j}\left[\mu\left(\frac{\partial v_i}{\partial x_j} + \frac{\partial v_j}{\partial x_i} - \frac{2}{3}\frac{\partial v_k}{\partial x_k}\delta_{ij}\right)\right] - \frac{\partial p}{\partial x_i} + \rho f_i \quad [1.28]$$

$$\frac{\partial(\rho e)}{\partial t} + \frac{\partial(\rho v_i e)}{\partial x_i} = \mu\left[\frac{\partial v_i}{\partial x_j}\left(\frac{\partial v_i}{\partial x_j} + \frac{\partial v_j}{\partial x_i}\right) - \frac{2}{3}\left(\frac{\partial v_i}{\partial x_i}\right)^2\right] - p\frac{\partial v_i}{\partial x_i} + \frac{\partial}{\partial x_i}\left(\kappa\frac{\partial T}{\partial x_i}\right) + \rho q \quad [1.29]$$

where ρ is the density of the fluid, e is the specific internal energy, the f_i are the external forces and q represents the heat sources.

For the heat flux h_i in the scalar equation [1.29], we used Fourier's law:

$$h_i = -\kappa_f \frac{\partial T}{\partial x_i} \qquad [1.30]$$

where κ_f is the thermal conductivity, the specific heat capacity is assumed to be constant, and the work performed by the pressure and friction forces is neglected. System [1.27]–[1.29] needs to be completed by two equations of state of the form:

$$p = p(\rho, T) \quad \text{and} \quad e = e(\rho, T) \qquad [1.31]$$

These relations determine the thermodynamic properties of the fluid. They are known as the thermal and caloric equations of state.

Usually, it is not necessary to solve the system of equations in its general form, and additional assumptions can be made to simplify the system. From a practical standpoint, the most useful assumptions are those of incompressibility and non-viscosity [SIG 11].

We will consider the Navier–Stokes equations describing the flow of incompressible Newtonian viscous fluids. This equations can be expressed in the fluid-filled domain Ω^f using the equations of conservation of momentum and the incompressibility condition:

$$\rho_f(\frac{\partial v}{\partial t} + v \cdot \nabla v) = -\nabla p + \mu_f \Delta v + \rho_f b_f \qquad [1.32]$$

$$\nabla \cdot v = 0 \qquad [1.33]$$

Equations [3.80] and [3.81] can be further expressed in dimensionless form by setting:

$$v_a = \frac{v}{V}, x_a = \frac{x}{L}, p_a = \frac{p - p_0}{\rho_0 V^2}, b_{fa} = \frac{b_f L}{V^2} \text{ and } t_a = \frac{tV}{L}$$

where x are the spatial coordinates, p_0 is a reference pressure, L and V are the characteristic length and velocity, respectively. Hence, by applying these relations to [3.80] and [3.81] and omitting the a index, we obtain the following equations:

$$\frac{\partial v}{\partial t} + v \cdot \nabla v = -\nabla p + \frac{1}{Re}\Delta v + b_f \qquad [1.34]$$

$$\nabla \cdot v = 0 \qquad [1.35]$$

We write Γ_D (or Γ_N) the boundary of Ω^f to which the Dirichlet boundary conditions (or surface stress conditions) are applied.

On Γ_D, writing $\bar{v}|_{\Gamma_D}$ for the Dirichlet data functions, we have

$$v = \bar{v}|_{\Gamma_D} \qquad [1.36]$$

By definition, the function t_f that describes the density of the contact forces on the fluid may be written as:

$$t_f = \sigma_f n_f \qquad [1.37]$$

where n_f is the outward unit normal vector on Γ_N.

On the edge Γ_N, we take boundary conditions [PIR 89, DEV 02] such that:

$$t_f = -pI_n + 2\mu_f d(v)n = \bar{t}$$

where $n = n_f$ and $\bar{t}$ is an imposed stress on Γ_N.

Equations describing the dynamics of elastic solids

In structural mechanics, in general, the goal is to determine the strain in a solid body induced by the action of various forces and to deduce the corresponding stresses, which are very important in a number of applications. There are many different laws that describe the various material properties. Combined with the equations describing the dynamics, these laws produce complex systems that enable us to find the strain (or the displacement).

Fundamental equations

In the field of structural dynamics, a distinction is usually made between linear and nonlinear models, which can either be geometric or physical. From a geometric perspective, linear problems are characterized by the linear relation between the stress and the strain, such that the following expression holds for the strain tensor ε_{ij}:

$$\varepsilon_{ij} = \frac{1}{2}\left(\frac{\partial u_i}{\partial x_j} + \frac{\partial u_j}{\partial x_i}\right) \qquad [1.38]$$

where u_i is the displacement vector. Physically, linear problems are based on a material law involving a linear relation between the strain and the stress. We will restrict ourselves to the case of linear elasticity, which is the most relevant in many industrial applications.

The theory of linear elasticity is geometrically and physically linear. As stated above, there is no need to distinguish between Eulerian and Lagrangian descriptions in the context of a geometrically linear theory.

The equations of linear elasticity theory are derived from linear stress–strain relations [1.38], which are conservation laws formulated in terms of the displacement [3.165] (known as equations of motion),

$$\rho\frac{D^2 u_i}{Dt^2} = \frac{\partial \sigma_{s_{ij}}}{\partial x_j} + \rho_s f_i \qquad [1.39]$$

and the assumption of linear elasticity in the material behavior, which is expressed by a constitutive equation of the form:

$$\sigma_{S_{ij}} = \lambda\varepsilon_{kk}\delta_{ij} + 2\mu\varepsilon_{ij} \qquad [1.40]$$

Finally, by eliminating ε_{ij} and T_{ij}, equations [1.38], [3.85] and [3.87] may be rewritten in the form of the following system of differential equations based on the displacement u_i:

$$\rho_S\frac{D^2 u_i}{Dt^2} = (\lambda+\mu)\frac{\partial^2 u_j}{\partial x_i \partial x_j} + \mu\frac{\partial^2 u_i}{\partial x_j \partial x_i} + \rho_s f_i \qquad [1.41]$$

These (unsteady) equations are known as the Navier–Cauchy equations for linear elasticity. In steady problems, we have that:

$$(\lambda + \mu)\frac{\partial^2 u_j}{\partial x_i \partial x_j} + \mu \frac{\partial^2 u_i}{\partial x_j \partial x_i} + \rho_S f_i = 0 \qquad [1.42]$$

In the solid, we denote Γ_D^s the edge with the Dirichlet boundary condition and Γ_N^s the section of the edge with surface stress conditions. Thus, on Γ_D^s, we can write that:

$$u = \bar{u}|_{\Gamma_D^s} \qquad [1.43]$$

Depending on the type of geometric model that will be considered later, we can fix internal conditions that complement the Dirichlet condition on the surface of the structure. Similarly, we denote t_s the density of the contact forces on Γ_N^s such that:

$$t_s = \sigma_s n_s \qquad [1.44]$$

where n_s is the outward unit normal vector from the solid.

Boundary conditions at the interface

The fluid–structure interaction describes the mechanical problem of the contact between an elastic solid body and a viscous fluid in motion. This essential step in our problem requires us to specify additional boundary conditions to describe the dialog between the fluid and the solid. Consider two domains, one fluid domain Ω^f and another solid domain Ω^s, in contact with each other along the interface Γ_I. We will use two conditions at the interface (Figure 1.14). On the one hand, we assume that the velocities are continuous with respect to time at the interface. This implies that

$$v = \dot{u} \quad \text{on} \quad \Gamma_I(t) \qquad [1.45]$$

On the other hand, we assume that the interface is in mechanical equilibrium

$$t_f + t_s = 0 \quad \text{on} \quad \Gamma_I(t) \qquad [1.46]$$

This second equation expresses the principle of the reciprocal action of forces at the interface.

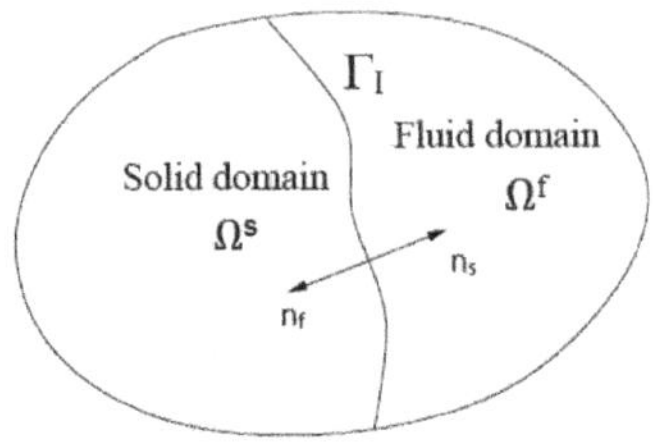

Figure 1.14. *Fluid–structure interface*

At the interface, the normal vectors are linked by the relation:

$$n = n_f = -n_s \qquad [1.47]$$

From [1.37], [1.44], [1.46] and [1.47], we can deduce that:

$$(\sigma_f - \sigma_s)n = 0 \quad \text{on} \quad \Gamma_I(t). \qquad [1.48]$$

2

Fluid–Structure Interaction with Ansys/Fluent

2.1. Presentation of Ansys

Ansys© is a sophisticated and comprehensive finite element software package. It is useful in many different areas of physics, such as structural and thermal analysis, Ansys fluids, FLOTRAN and CFD. It can also be used to analyze coupled domains involving one or multiple types of these different kinds of physics. Ansys is an integrated program that allows all of these operations to be performed via a graphical user interface. Model creation and execution, as well as results postprocessing, can be done without leaving the Ansys environment. Because of its extremely fast, accurate and reliable numerical simulation tools, Ansys can be used to obtain accurate solutions to the problems encountered during the design and development of products. It allows organizations from all activity sectors to predict the performance of their products after market launch. Ansys is a general purpose software package for simulating interactions in all disciplines of physics, structural and vibrational analysis, and fluid dynamics, as well as heat transfer and electromagnetic analysis in engineering. Ansys allows designs to be tested in a virtual environment before a product prototype is manufactured. With Ansys, weaknesses can be identified and improved, the product lifetime can be computed, and likely issues can be predicted by conducting three-dimensional simulations in virtual environments.

Ansys is used by the following non-exhaustive list of industries: aerospace, automotive, biomedical, bridge and building engineering, electronics and appliances, heavy equipment and machinery, etc. It is important to note that the Computer-Aided Design (CAD) aspect is not restricted to the finite element model itself, which is just a way of specifying to Ansys the positions of the nodes and the elements required to implement the finite element method. CAD consists of finding a mesh in order to create nodes and elements. The Finite Element Method (FEM) subdivides a structure into a numbers of elements (pieces of the structure), gives a simple description of the behavior of each individual element and then reconnects these elements together once again. This process is described by a set of simultaneous algebraic equations.

The key products associated with Ansys include:

– *Ansys Structural*: this product allows mechanical simulations to be performed in order to compute structures. Its primary features are as follows:

- static analysis;

- modal analysis;

- harmonic analysis (forced response);

- temporal analysis;

- handling of various different nonlinear situations (contact, material plasticity, large displacements or deformations).

– *Ansys Mechanical*: this product has the same features as Ansys Structural with the addition of a thermal solver with radiation modeling.

– *Ansys CFX and Fluent:* these two software programs can be used to perform simulations in fluid mechanics. They bear the names of the companies that developed them, which were acquired by Ansys Inc. in February 2003 and February 2006, respectively. Fluent is a solver: it does not include a mesher (the mesh must be constructed using meshing software, such as Gambit, which is also published by Ansys). The parameters of the model are configured using a graphical interface. A scripting interface is also available to automate the computation procedures. One of the advantages of this general purpose simulation software is that it has a relatively large number of models that can be implemented to solve questions in fluid mechanics: diphasic flows (miscible, immiscible, cavitation, solidification), turbulence (LES, KE, Kw, SA, Reynolds stress), combustion (premixed and non-premixed), particle transport, flows through porous media, mobile and dynamic meshes with mesh reconstruction and many others. The temporal and spatial numerical schemes of these models can be adjusted in order to improve their convergence. Fluent supports parallel processing and can take advantage of multiprocessor systems both on a single machine and over a network (cluster, dual core, multi-CPU platforms):

- *Gambit*: meshing software published by Ansys since 2006 (previously published by Fluent). This mesher allows custom geometries to be created with high levels of freedom and accuracy, or alternatively imported from a CAD file. It can also automatically mesh surfaces and volumes in parallel to the specification of boundary conditions. Gambit is often viewed as the gold standard for meshers by Fluent users.

– *Ansys AUTODYN and Ansys LS-DYNA*: these software packages have solvers that use explicit formulations for the equations to be solved, unlike the products listed above. Their scope of application is confined to models whose mechanics involve very large deformations.

– *Ansys Electromagnetics and Ansoft*: this product allows models involving electromagnetic phenomena to be solved.

– *Ansys Multiphysics*: this product combines all Ansys tools for implicit numerical simulations.

Two software environments can be used to implement Ansys code:

– *Ansys Classic*: chronologically, this was the first software solution offered by the developers. It is designed for constructing finite element models with simple geometries that can easily be assembled from basic operations. Within this environment, the user directly builds finite element models using the scripting language i.e. Ansys Parametric Design Language (APDL). Ansys Classic is therefore oriented toward users with experience in numerical simulations.

– *Ansys Workbench*: this platform takes a different approach to model building by reusing the original Ansys code. It is particularly suitable for handling problems with complex geometries (many objects with pieces) and is accessible to users without experience in performing computations. In this environment, the user essentially works on the geometry of the model and not the model itself. The platform therefore fulfills the task of converting the commands specified by the user into Ansys code before launching the solving procedure. The finite element models generated in this way may, however, still be manipulated by inserting custom commands into the Ansys code.

2.2. Coupling with Ansys

Analyzing a coupled domain requires a combination of different analysis components from different disciplines in engineering (physical domains), which interact together to develop the solution to a global engineering problem. To reflect this, we often describe coupling analysis as multiphysics analysis. When the inputs of one part of the analysis depend on the results of another, we say that these analysis components are coupled.

Some analysis components have unidirectional couplings. For example, in thermal stress problems, the temperature field places thermal constraints on the structure domain, but the structural deformations generally do not influence the temperature distribution. Therefore, it is not necessary to iterate between the solutions of both domains. More complex situations can involve bidirectional couplings. Piezoelectric analysis, for example, studies the interaction between the electrical and structural fields: it allows the voltage distribution to be found as a result of the applied displacement or vice versa. In fluid–structure interaction (FSI) problems, the fluid pressure causes a structural deformation, which in turn modifies the solution for the fluid. In order to achieve convergence in this problem, it will be necessary to iterate between both physical domains. The coupling between the domains can be described directly or via load transfer. This coupling might be complicated, since each of the fields may have been solved using different types of analysis during the simulation. Examples of applications that require coupling include pressure-driven devices (thermal stress analysis), obstacles in fluid flows (FSI), induction heating (magnetic–thermal analysis), ultrasonic transducers (piezoelectric analysis), magnetic forming (magneto-structural analysis) and microelectromechanical systems.

2.2.1. *Types of coupling analysis*

The procedure for coupling analysis depends on the coupled domains, but two distinct methods can be identified: the direct method and the load transfer method.

The direct method

The direct method generally involves a single analysis that uses coupled field elements containing all necessary degrees of freedom. The coupling is carried out by calculating the load matrices or vectors that group together all of the necessary terms. One example of coupling using the direct method is piezoelectric analysis using the PLANE223, SOLID226 or SOLID227 elements [ANS 12b]. Another example is FLOTRAN analysis, which uses FLOTRAN elements [ANS 12b].

The load transfer method

Load transfer methods involve two or more types of analysis, each associated with a different domain. Coupling between two domains is achieved by applying the results of one analysis as loads in another analysis. There are different kinds of load transfer analysis.

Load transfer coupling Workbench: system coupling

The coupling can be performed using a system of components called "system coupling" in Workbench. More precisely, we can implement the analysis of unidirectional or bidirectional FSIs by connecting the "system coupling" component to external mechanical or fluid data systems.

Load transfer coupling – Ansys multifield solver

Ansys multifield solver, available for a wide class of coupling problems, is an automated tool for solving load transfer problems. It provides a robust, accurate and easy-to-use tool for solving physics problems involving load transfer couplings. The solver automatically transfers the coupled loads to each of the different meshes. It can be used for static, harmonic and transient analysis, depending on the physical requirements, and arbitrarily many fields can be solved sequentially.

There are two versions of the Ansys multifield solver, designed for different applications, and each of which offers specific advantages and different procedures:

– *MFS–single code*: ansys multifield solver basic is used if the simulation involves small models and if all physical domains are contained within a single executable product (for example Ansys multiphysics).

– *MFS solver*: single code uses iterative coupling and solves each physical domain sequentially, solving each matrix equation separately. The solver iterates between each physical domain until the loads transferred through the physical interfaces converge.

Load transfer coupling – physics file

When using a physics file, we need to explicitly transfer the loads by using one environment for each physical domain. One example of this kind of analysis is sequential thermal stress analysis, in which the nodal temperatures from thermal analysis are applied as loads in the subsequent stress analysis. This method is based on a single finite element mesh shared by all environments, requiring physics files to be created to define the environment of each domain; these files configure the database and prepare the shared mesh for each given physics simulation. The general procedure is as follows:

– read the first physics environment and solve it;

– read the next environment, indicating the loads to be transferred;

– solve the second physics environment.

The LDREAD command allows different physics environments to be connected together and enables the data of the results specified by the first physics environment to be applied as loads in the solution of the next environment through a similar interface.

Load transfer coupling – unidirectional load transfer

Another way of implementing a fluid solid coupling is by unidirectional load transfer, so that the results of the fluid analysis do not significantly affect the solid loads and vice versa. The loads can be transferred unidirectionally from Ansys multiphysics to the CFX fluid, or in the opposite direction, so that the load transfer occurs externally from the analysis.

2.3. Example of fluid–structure interaction using the "physics" environment

2.3.1. *Fluid in motion*

The example in this section illustrates a fluid–structure interaction problem in a steady state, with an application of large nonlinear structural deformations coupled with a fluid domain. Water flows through a pipe with a rubber seal at an inlet velocity of 0.35 m/s. The objective of the problem is to find the deformation of the seal under the pressure of the water flow in equilibrium conditions.

The pressure of the fluid causes the seal to experience a deformation. This deformation may be sufficiently large as to affect the flow. In this case, the example defines a small region of fluid around the watertight seal, which is used by the physics environment of the fluid. By performing structural analysis in the structure region, we find the displacements of the seal elements required to transform the small region around the seal. We will later use the morph mesh for fluid analysis. The fluid analysis uses null-type elements for the seal, and the structure analysis uses null-type elements for the fluid.

To compute the finite elements using the Ansys© code, the fluid and the structure are meshed with rectangular elements of four nodes, FLUID141 and

PLANE182, respectively. The chosen meshes are compatible at the fluid–structure interface, where the nodes of the two domains coincide. Figure 2.1 shows the model of the coupled problem and its finite elements discretization.

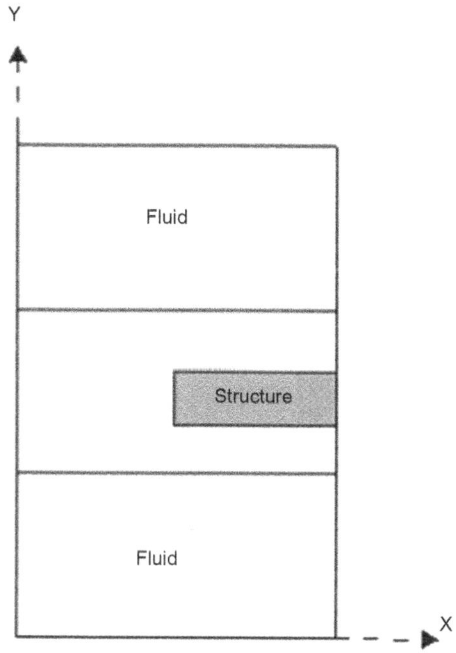

Figure 2.1. *Structure within a fluid. For a color version of this figure, see www.iste.co.uk/elhami/interactions.zip*

The initial FLOTRAN analysis needs to be repeated until convergence is achieved. Subsequent instances of analysis will reuse this solution, and so should not require as many global iterations to converge. Subsequent instances of structural analysis will also be restarted.

After each structural analysis, the mesh transformation is calculated to move the nodes within the small region of fluid around the watertight seal so that they are consistent with the displacement of the structure. The updated positions of these nodes will be taken into account in subsequent analysis steps for the flow. In order to find an adequate solution for the structure, all of the nodes must be returned to their original positions before the updated pressure values of the fluid flow are applied.

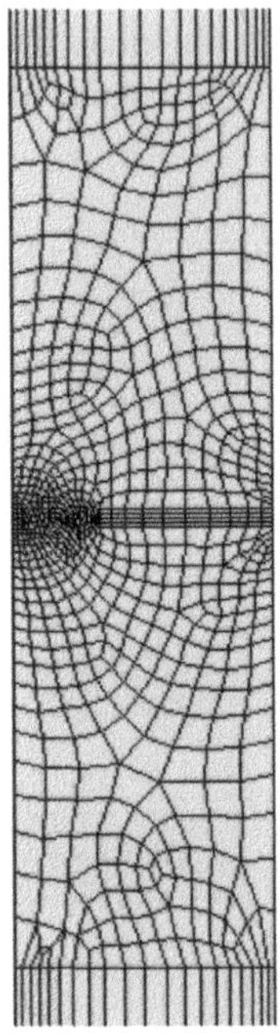

Figure 2.2. *Finite element mesh of the coupling. For a color version of this figure, see www.iste.co.uk/elhami/interactions.zip*

Results

The fluid–structure solving loop was executed until the convergence criteria were met. The convergence tolerance was chosen to be 0.5. In the first analysis, 400 global iterations were enough to allow the FLOTRAN solution to converge. In the fluid–structure interaction loop, the number of iterations was fixed at 100 for the remaining FLOTRAN routines.

Figures 2.3–2.5 show the results of the structural displacement, the pressure, the flow velocity and the current lines near the deformed seal.

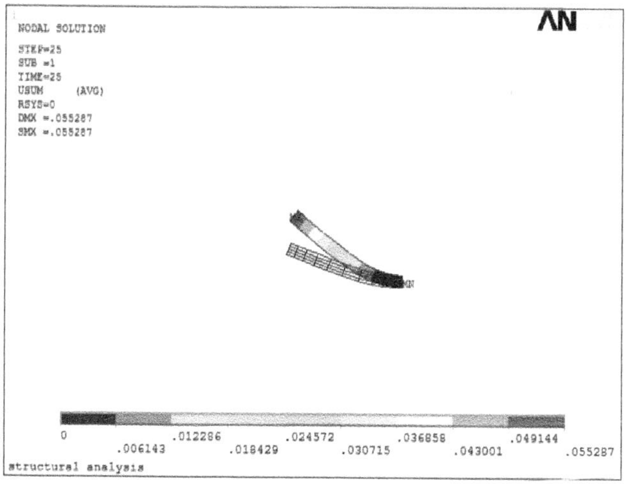

Figure 2.3. *Structural displacement. For a color version of this figure, see www.iste.co.uk/elhami/interactions.zip*

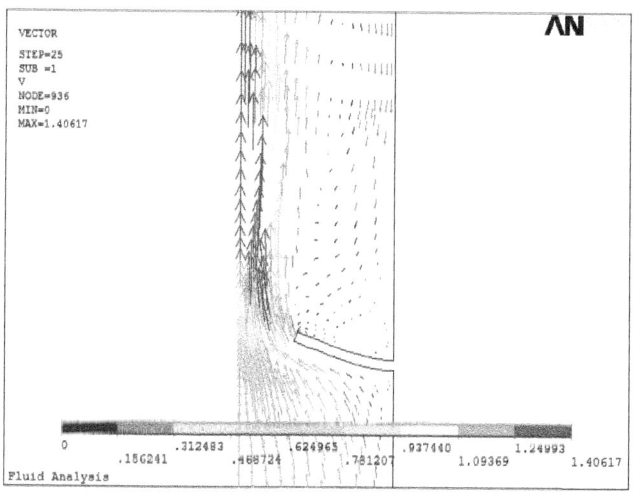

Figure 2.4. *Fluid pressure. For a color version of this figure, see www.iste.co.uk/elhami/interactions.zip*

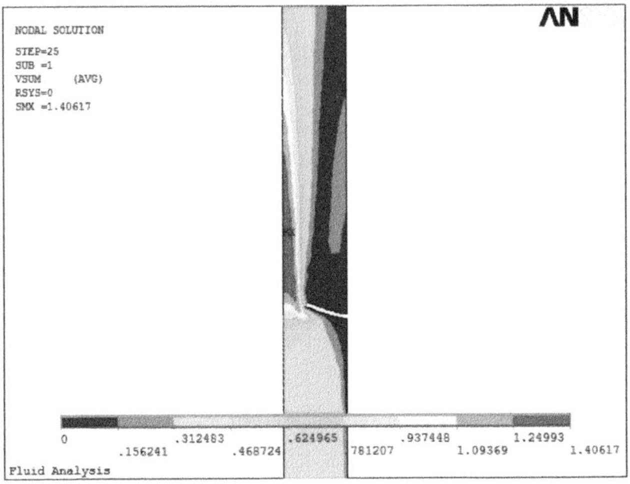

Figure 2.5. *Flow velocity. For a color version of this figure, see www.iste.co.uk/elhami/interactions.zip*

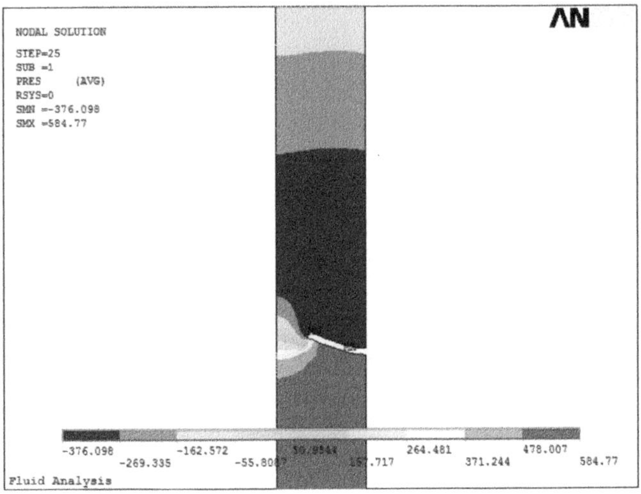

Figure 2.6. *Current lines. For a color version of this figure, see www.iste.co.uk/elhami/interactions.zip*

This is implemented by the following program written in APDL:

```
/prep7 \\
ET,1,141\\
ET,2,182,\\
!!!!!!!
!!!!!!!

yent = 0.0
dyen = 1.0
ysf1 = yent+dyen dsf1 = 0.5
ygas = ysf1+dsf1
dg = 0.02 dg2=dg/2.
ytg = ygas+dg dsf2 = 0.5
ysf2 = ytg + dsf2

dyex =  6.0 x = 0.
dgasr =.20 piper = 0.3
xrgap = piper-dgasr !! !!!

!Create the geometry

!!
rect,xrgap,piper,ygas,ytg
rect,x,piper,ysf1,ysf2
rect,x,piper,yent,ysf1
rect,x,piper,ysf2,ysf2+dyex
aovlap,all

! Mesh parameters

ngap = 10
ngas = 10
rgas = -2
nflu = ngap+ngas
raflu = -3
nenty =8
raent =5
nfl1 = 20
nthgas = 4
nfl2 = 3
next = 30
```

```
rext = 6
rafls = 12
lesize,1,,,ngas,rgas
lesize,3,,,ngas,rgas
nfl11= nfl1*2+9
lsel,s,,,2,4,2
lesize,all,,,nthgas
alls
lesize,5,,,nflu,raflu
lesize,7,,,nflu,raflu
lesize,9,,,nflu,raflu
lesize,15,,,nflu,raflu
lesize,18,,,nenty,1./raent
lesize,17,,,nenty,1./raent
lesize,21,,,nfl1,rafls
lesize,8,,,nfl11,-1./(rafls+3)
lesize,22,,,nfl1,rafls
lesize,19,,,next,rext
lesize,20,,,next,rext

!!! AATT,MAT,REAL,TYPE
asel,s,,,1
aatt,2,2,2
asel,s,,,7
cm,area2,area
alist

asel,a,,,5,6
aatt,1,1,1    ! Fluid Surface (material 1)
alls

eshape,2
asel,u,,,7
amesh,all
eshape,0
asel,s,,,7
amesh,all

!!  2.! Create the physics environment for the fluid

et,1,141 et,2,0 vin=3.5e-1 !!          CFD Flow condition
flda,solu,flow,1
```

```
flda,solu,turb,1 flda,iter,exec,400
flda,outp,sumf,10 !! CFD Fluid properties
flda,prot,dens,constant
flda,prot,visc,constant
flda,nomi,dens,1000.
flda,nomi,visc,4.6E-4 flda,conv,pres,1.E-8 !!
       CFD application of boundary conditions
lsel,s,,,8,17,9 lsel,a,,,20 dl,all,,vx,0.,1
lsel,s,,,9 dl,all,,vx,0.,1 dl,all,,vy,vin,1 lsel,s,,,2
lsel,a,,,18,19 lsel,a,,,21,22 dl,all,,vx,0.,1 dl,all,,
      vy,0.,1
lsel,s,,,1,3,2 dl,all,,vx,0.,1 dl,all,,vy,0.,1 lsel,s,,,15
dl,15,,pres,0.,1

alls
/title,Fluid Analysis
physics,write,fluid,fluid
!!

!!!!! 3. Create the physics environment for the structure
!!
physics,clear
!SOLCONTROL, , , NOPL,
et,1,0
et,2,182
keyopt,2,3,2
keyopt,2,6,1
keyopt,2,1,2
!mp,nuxy,2,0.49967
!tb,mooney,2
!tbdata,1,0.293E+6  !  Mooney-Rivlin Constants
!tbdata,2,0.177E+6  !     "      "         "
tb,hyper,2,,2,mooney
tbdata,1,0.293E+6,0.177E+6,
      (1.0-2.0*0.49967)/(0.293E+6+0.177E+6)

lsel,s,,,2
nsll,,1
d,all,ux,0.
d,all,uy,0.
alls
```

```
/title,structural analysis
finish
/solu
antype,static
nlgeom,on
cnvtol,f,,,,-1
physics,write,struc,struc
physics,clear

!!! Fluid-Structure Interaction
FSSOLV,Fluid,Structure,2,AIRE2,,,,,25,300,1,1
```

2.3.2. *Stagnant fluid*

In this section, we will present a steady-state fluid–structure interaction problem with small, linear structural deformations. The example is that of an elastic plate with Young's modulus $E = 2.11e11$ Pa, density $\rho_s = 7,860$ kg.m^3 and Poisson coefficient $\nu = 0.3$, fully immersed in a compressible fluid at rest (with density $\rho_f = 1,000$ kg.m^3 and speed of sound $c = 1,500$ ms^1). The objective of the problem is to determine the natural frequencies of the structure coupled with the water and the effect of the coupling on the vibrations of the system, without model reduction. The structure is assumed to be mounted from the top end, and is not subject to a load, so that the fluid may be assumed to be immobile at the edges and the pressure is also zero at the edges except on the top end, which is assumed to be free. To compute the finite elements using Ansys©, the fluid and the structure are meshed with rectangular elements of four nodes, FLUID29 and PLANE182, respectively. The chosen meshes are compatible at the fluid–structure interface, where the nodes of the two domains coincide, and which has 600 elements: 100 for the structure and 500 for the fluid. Each element has four nodes, and each node has three degrees of freedom U_x and U_y (corresponding to displacement in the x and y directions) and P, the pressure at the interface. Figure 2.7 shows the model of the coupled problem and its finite element discretization.

The first acoustic natural modes are shown in Figures 2.8–2.12.

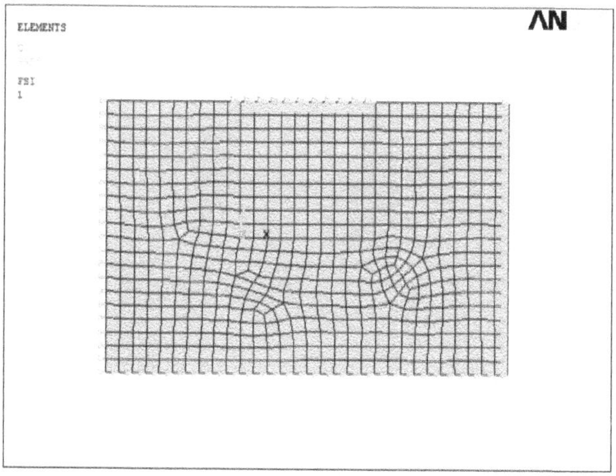

Figure 2.7. *Discretization of the coupled model. For a color version of this figure, see www.iste.co.uk/elhami/interactions.zip*

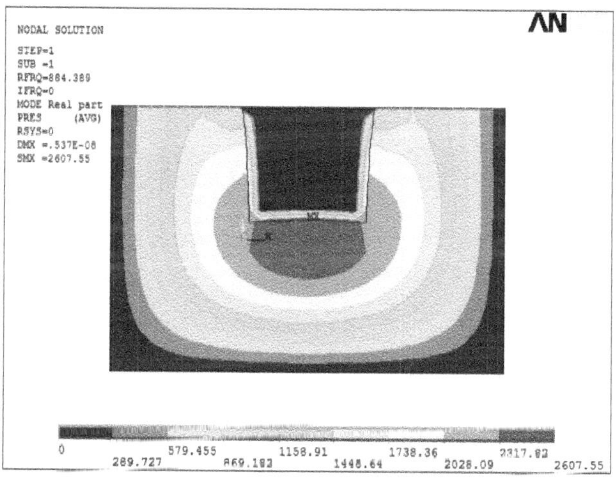

Figure 2.8. *First structural mode. For a color version of this figure, see www.iste.co.uk/elhami/interactions.zip*

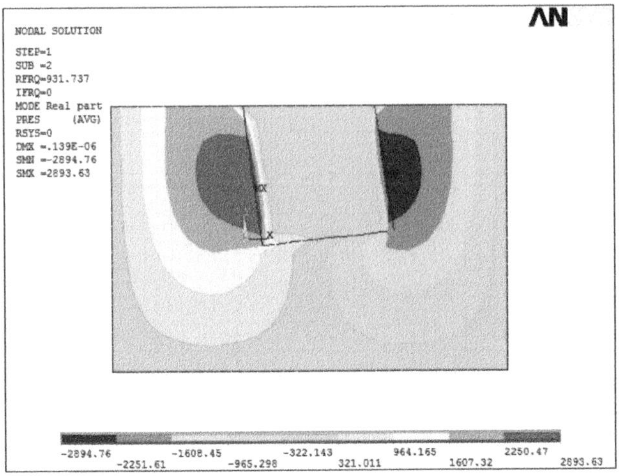

Figure 2.9. *Second structural mode. For a color version of this figure, see www.iste.co.uk/elhami/interactions.zip*

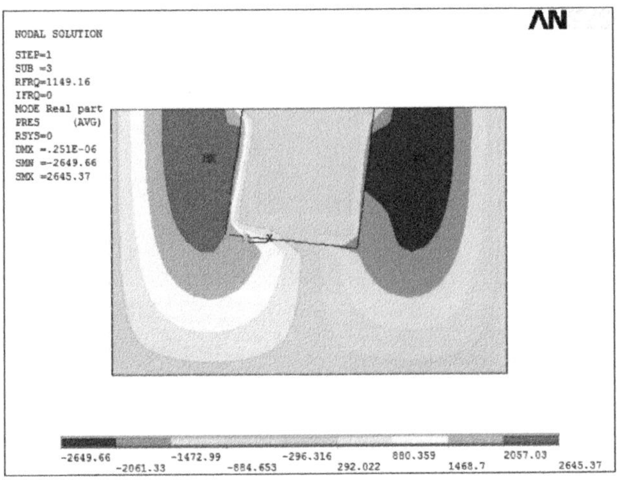

Figure 2.10. *Third structural mode. For a color version of this figure, see www.iste.co.uk/elhami/interactions.zip*

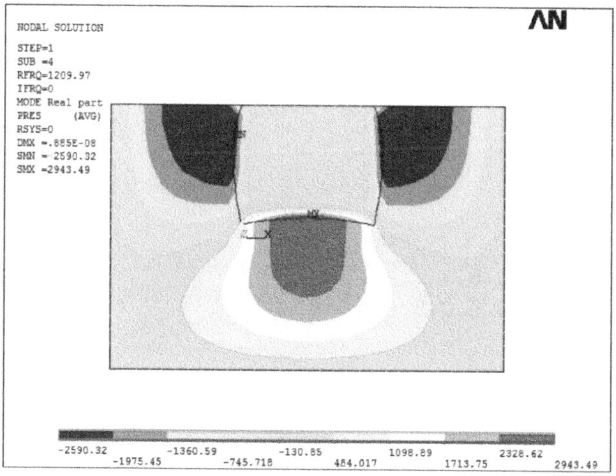

Figure 2.11. *Fourth structural mode. For a color version of this figure, see www.iste.co.uk/elhami/interactions.zip*

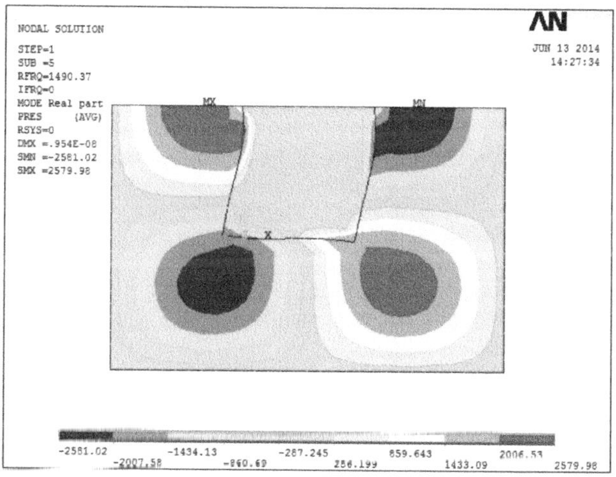

Figure 2.12. *Fifth structural mode. For a color version of this figure, see www.iste.co.uk/elhami/interactions.zip*

This is implemented by the following program written in APDL:

```
/PREP7
/UNITS,SI

Et,1,PLANE182
Et,2,FLUID29

RECTNG,0,0.5,0,0.5,
RECTNG,-0.5,1,-0.5,0.5,
aovlap,all

! material properties
MP,EX,1,2.11e11
MP,DENS,1,7860
MP,NUXY,1,0.3
MP,DENS,2,1000
MP,SONC,2,1500

! select, assign attribute to and mesh area 1

ASEL,S,AREA,,1
AATT,1,1,1,0
LESIZE,1,,,10,1
LESIZE,3,,,10,1
LESIZE,2,,,10,1
LESIZE,4,,,10,1
MSHKEY,1
MSHAPE,0,2D
AMESH,1

! select, assign attribute to and mesh area

ASEL,S,AREA,,3
AATT,2,2,2,0
LESIZE,8,,,20,1
LESIZE,5,,,30,1
LESIZE,6,,,20
LESIZE,9,,,10
LESIZE,10,,,10
LESIZE,2,,,10
LESIZE,4,,,10
```

```
LESIZE,1,,,10
MSHKEY,0
MSHAPE,0,2D
AMESH,3


! flag interface as fluid-structure interface

LSEL,S,,,1
LSEL,A,,,2
LSEL,A,,,4
NSLL,S,1
esel,s,type,,2
sf,all,fsi,1
nsel,all
esel,all
allsel,all

!!!!Boundary conditions and loads

LSEL,s,,,5
LSEL,a,,,6
LSEL,a,,,8
NSLL,s,1
D,all,PRES,
allsel,all

LSEL,s,,,3
NSLL,s,1
DL,ALL,,ALL,
allsel,all

/SOLU
ANTYPE,MODAL
MODOPT,UNSYM,5
MXPAND,5
SOLVE
FINISH
```

2.4. Example of interaction using Fluent

This example presents a fluid–structure interaction problem in two parts. The first part involves creating and solving an air flow with Fluent, and passing on the forces imposed by the fluid in order to calculate the deformation of the structure.

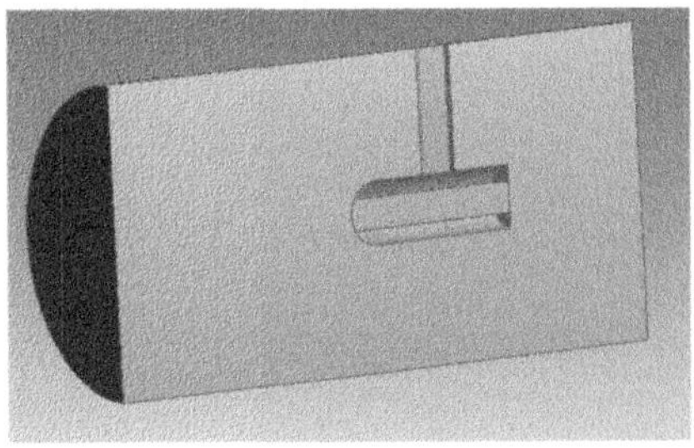

Figure 2.13. *Air flow. For a color version of this figure, see www.iste.co.uk/elhami/interactions.zip*

In Figure 2.14, blue is the internal speed with value 30 m/s, red is the external pressure of 0 Pa, white is the wall, and yellow shows the symmetry of the problem.

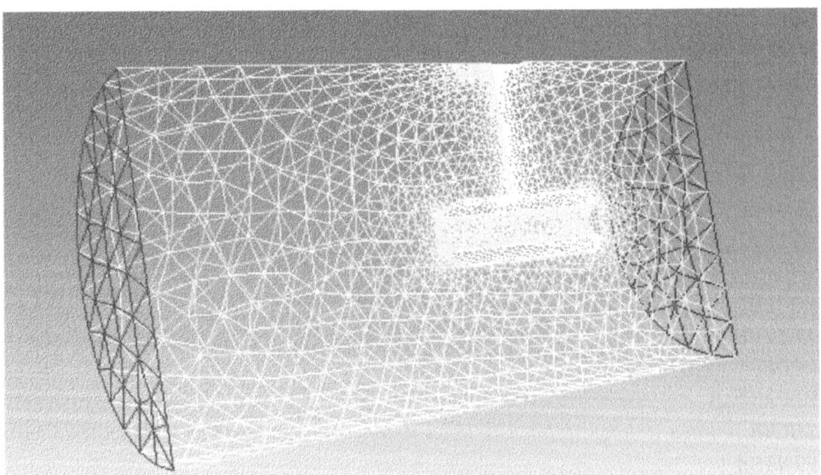

Figure 2.14. *Mesh relative to the air flow. For a color version of this figure, see www.iste.co.uk/elhami/interactions.zip*

The mesh consists of 13, 133 nodes and 61, 631 elements. We also consider a turbulent flow solved by the K-epsilon method, and the turbulent kinetic energy is solved using the first-order upwind method.

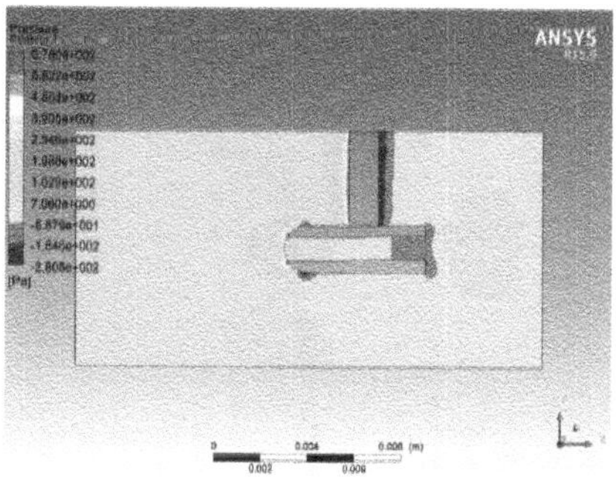

Figure 2.15. *Pressure contour. For a color version of this figure, see www.iste.co.uk/elhami/interactions.zip*

Next, we analyze the behavior of the structure, using the results from Fluent to define the pressure. The structure is fixed from the top, but its other sides are free.

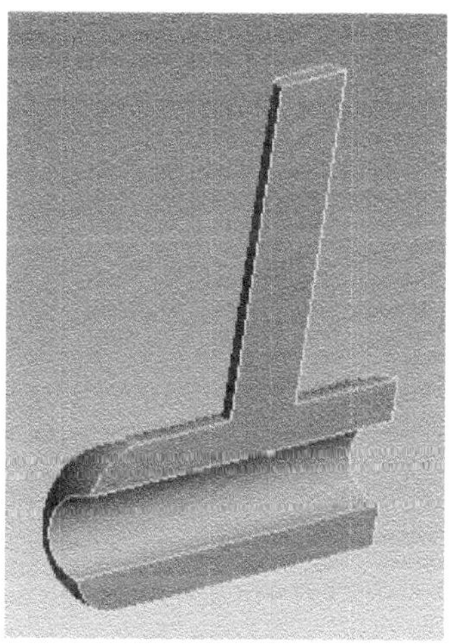

Figure 2.16. *Model of the structure. For a color version of this figure, see www.iste.co.uk/elhami/interactions.zip*

The mesh uses 3D tetrahedra, and has 698 nodes and 297 elements (see Figure 2.17). We import the fluid pressure onto the interfaces of the fluid–structure interaction (see Figure 2.18). We solve the problem using static analysis to calculate the total displacement of the structure (see Figure 2.19).

Figure 2.17. *Mesh of the structure. For a color version of this figure, see www.iste.co.uk/elhami/interactions.zip*

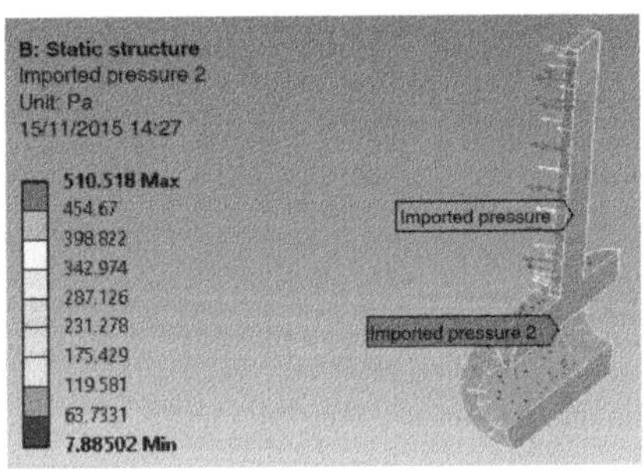

Figure 2.18. *Behavior of the structure. For a color version of this figure, see www.iste.co.uk/elhami/interactions.zip*

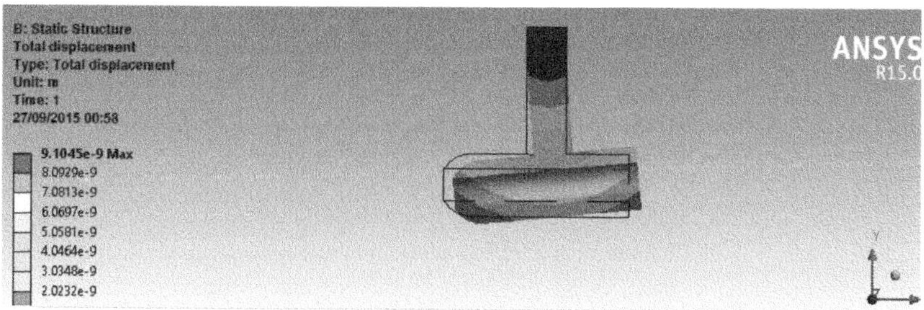

Figure 2.19. *Total displacement of the structure. For a color version of this figure, see www.iste.co.uk/elhami/interactions.zip*

3

Vibroacoustics

3.1. Introduction

Fluid structure interaction involves studying the behavior of a solid in contact with a fluid in situations where the response of the solid is strongly affected by the action of the fluid. These phenomena arise everywhere, and can sometimes explain why things work the way they do, or do not work, as the case may be. Vibrations affect the integrity of structures and must be predicted in order to prevent the system from experiencing accelerated wear or even collapsing if the vibrations exceed a certain threshold. Currently, there are several interesting fluid–structure problems, for example in environmental engineering with the transport of toxic products as fluids and flows around turbine blades, in the automotive industry, aeronautics and aeroacoustics, the transportation industry and the shipbuilding industry, where the responses exhibited by coupled systems remain a poorly understood and delicate topic. Studying these phenomena requires us to perform vibroacoustic analysis, which has inspired specialized lines of scientific research into the behavior of structures at medium and high frequencies.

Our choice of strategy for tackling the coupled problem will depend on the medium that we wish to emphasize (fluid or structure). Information is transferred from the secondary medium to the primary medium. In the applications considered here, the primary medium will usually be the structure, and the fluid variables will be eliminated in favor of the structure variables. It is therefore clear just how important it is to prepare reliable numerical models for any project in advance to allow these kinds of behavior to be predicted.

The objective of this chapter is to present a deterministic model of vibroacoustic systems and state the discretized equations of the problem. We will begin by briefly recalling the general equations describing the behavior of the fluids and structures, which we will analyze with the classical assumptions of small movements and small deformations. The problem is formulated in three parts: one for the structure, one for the acoustic cavity and one for the coupled vibroacoustic system. In each part, we present the boundary value problem, its weak formulation and its discretization obtained by the finite element method.

To solve the coupled problem, we will use a coupled numerical method, finite elements/finite elements or finite elements/finite volumes, that is well suited to solving vibroacoustics problems and analytical methods that only apply to certain very particular types of problem, namely almost exclusively problems with rotational symmetry. We will illustrate each of these methods in more depth when we study the dynamic behavior of a two-dimensional (2D) and three-dimensional(3D) plate and a beam immersed in a compressible fluid.

3.2. Equations of the acoustic and structure problems

3.2.1. *Equation of the acoustic problem*

The fluid is analyzed in terms of Eulerian variables and is described by two local equations:

– Continuity equation or conservation of mass:

$$\frac{\partial \rho_f}{\partial t} + \nabla \rho_f . v = 0 \ \text{ in } \Omega_f \tag{3.1}$$

where ρ_f is the fluid density and v is the velocity vector of the particles.

– Equation of conservation of momentum:

$$\rho_f \left(\frac{\partial v}{\partial t} + \frac{1}{2} \nabla v \wedge v \right) = f + \nabla \sigma \quad \text{in } \Omega_f \tag{3.2}$$

The stress tensor σ is expressed as a function of the pressure p, the viscosity and the infinitesimal strain tensor $D = \frac{1}{2} \left(\nabla v + \nabla^t v \right)$:

$$\sigma = pI + 2\mu D \tag{3.3}$$

We make the following assumptions:

– perfect fluid, thus with zero viscosity: $\sigma = pI$;

– irrotational motion rot $v = 0$, therefore there exists a velocity potential Φ such that $v = \nabla\Phi$;

– small fluid motion hypothesis, therefore, neglecting the v^2 term compared to the other terms, we have: $p = -\rho_f \frac{\partial \Phi}{\partial t}$;

– compressible fluid: we will assume adiabatic behavior, which gives the Helmholtz equation (wave equation):

$$\Delta p - \frac{1}{c^2}\frac{\partial^2 p}{\partial t^2} = 0 \quad \text{in } \Omega_f \qquad [3.4]$$

and therefore:

$$\Delta \Phi - \frac{1}{c^2}\frac{\partial^2 \Phi}{\partial t^2} = 0; \quad \text{in } \Omega_f \qquad [3.5]$$

where c is the speed of sound in the fluid.

3.2.2. *Boundary conditions of the acoustic problem*

– Slip condition on the walls of the solid:

This is obtained by stating that the fluid particles cannot penetrate the walls of the solid, that is to say that the normal velocities of the fluid and solid particles are continuous:

$$v_s.n = v.n = \nabla\Phi \quad \text{on } \Sigma \qquad [3.6]$$

which implies that:

$$v_s.n = \frac{\partial \Phi}{\partial n} \quad \text{on } \Sigma \qquad [3.7]$$

This condition may also be written in terms of the pressure as follows:

$$\frac{\partial p}{\partial n} + \rho_f.\ddot{u} = 0 \quad \text{on } \Sigma \qquad [3.8]$$

where $n = n_s = -n_f$, n_s is the outward normal of the structure domain, n_f is the outward normal of the fluid domain, $\ddot{u}$ is the normal acceleration of the structure and Σ is the fluid–structure interface.

– Free surface condition:

$$p = 0 \text{ and } \Phi = 0 \qquad [3.9]$$

3.2.3. *Equation of the structure problem*

We work within the framework of linear elasticity with the assumption of small perturbations. The local equations are obtained in a Galilean coordinate system by stating that at each time t, on any section Ω_s of a material system, the dynamic torsor of Ω_s is equal to the torsor of the external forces exerted upon Ω_s, and the equilibrium equations may be written as:

$$\nabla\sigma^s + f = \rho_s\ddot{u}; \quad in \;\; \Omega_s \qquad [3.10]$$

where σ^s is the symmetric Cauchy stress tensor, f is the force distribution vector, $\ddot{u}$ is the acceleration at the point M and u is the displacement field.

The strain vector associated with the infinitesimal strain tensor satisfies:

$$\epsilon^s = \langle \epsilon_{xx} \;\; \epsilon_{yy} \;\; \epsilon_{zz} \;\; 2\epsilon_{xy} \;\; 2\epsilon_{yz} \;\; 2\epsilon_{zx} \rangle \qquad [3.11]$$

the stress–strain relation (generalized Hooke's law) may be written as:

$$\sigma = D\epsilon \qquad [3.12]$$

where D is the elasticity matrix and σ is the stress tensor associated with the Cauchy stress tensor σ^s with components:

$$\sigma^s = \langle \sigma_{xx} \;\; \sigma_{yy} \;\; \sigma_{zz} \;\; \sigma_{xy} \;\; \sigma_{yz} \;\; \sigma_{zx} \rangle \qquad [3.13]$$

3.2.4. *Boundary conditions of the structure problem*

Two types of boundary conditions are applied to the boundary of the structure domain:

– natural boundary conditions:

$$\sigma^s.n = T_d \quad \text{on} \quad \Sigma_{s\sigma} \qquad [3.14]$$

– geometric (or kinematic) boundary conditions:

$$u = u_0 \quad \text{on} \quad \Gamma_0 \qquad [3.15]$$

where $\Sigma_{s\sigma} \cap \Gamma_0 = \phi$ or $\Sigma_{s\sigma} \cap \Gamma_0 = \Sigma_s$.

3.3. Vibroacoustic problem

There are two approaches to solving coupled fluid–structure problems. The first is to simultaneously solve all coupled equations by combining the degrees of freedom associated with each model into a single matrix. This type of coupling is described as strong coupling. The second approach is to solve the equations sequentially, fixing the parameter set of each model and iterating the other model. This is described as weak coupling. The advantage of the first method is that it is likely to produce more accurate results at the cost of excessively high computation times. The second method can benefit from the fast computation times offered by specialized solvers, but has a higher risk of instability and produces less accurate results [AXI 01, SOU 09]. Several formulations for modeling this problem have been suggested in the literature. These formulations include:

– classical variational formulation in terms of the displacement u for the structure and the pressure p for the fluid (the advantage of this formulation is that there is one single unknown per node, however it leads to a non-symmetric matrix system);

– variational formulation in terms of the displacement u for the structure and the average pressure p_m for the fluid (the advantage of this numerical model is that the fluid cavity is simulated at surfaces, and so these surfaces can be meshed using 2D finite elements). However, this model is only suitable when the cavity is thin relative to the acoustic wavelength;

– variational formulation in terms of (u, γ, p), which is based on a description of the structure in terms of two vector fields (the displacement u and the acceleration γ of the structure for harmonic problems with respect to time $\gamma = -\omega^2.u$) and a description of the fluid in terms of a single scalar field (the pressure p of the fluid cavity). The advantage of this description is that it results in a symmetric system. However, it has the disadvantage that this system is large;

– variational formulation in terms of (u, p, ϕ), which describes the structure by its displacement field u and describes the fluid by two scalar fields (the pressure p and the velocity potential ϕ). The advantage of this formulation is that, after discretization by finite elements, it results in a large symmetric matrix system;

– variational formulation in terms of (u, p, ϕ), which describes the structure by its displacement field u and describes the fluid by two scalar fields (the pressure p and the displacement potential ϕ). This formulation results in a symmetric matrix system containing all degrees of freedom of the problem.

In this section, we present a numerical model for calculating variables in a coupled fluid–structure system based on the first and final two of the formulations listed above. This model requires us to perform the following different steps:

– write the equations of motion of the structure coupled with a fluid cavity to determine the variational form associated with the coupled elastoacoustic system;

– apply the finite element method to discretize the structure and the fluid cavity. This discretization gives us the matrix equation that determines the motion of the elastoacoustic system.

3.3.1. *Problem statement*

The vibroacoustic problem consists of a structure coupled with an acoustic cavity. The system is modeled by a structure domain Ω_s made from an elastic material and a domain Ω_f containing an acoustic medium modeled as a perfect, weightless, evolution-compressible, isentropic fluid. We denote by $\Sigma = \Sigma_f \cap \Sigma_s$ their shared surface.

The system is formulated in the Cartesian coordinate system (O, x_1, x_2, x_3). The outward normal to Ω_s is written as $n_s = (n_{s_1}, n_{s_2}, n_{s_3})$, and the outward normal to Ω_f is written as $n_f = (n_{f_1}, n_{f_2}, n_{f_3})$. The normal to Σ directed outwards from Ω_f is written as n_s.

The displacement field u at each point in the domain Ω_s at a given frequency ω is $u(x, \omega) = (u_1(x, \omega), u_2(x, \omega), u_3(x, \omega))$. The structure is fixed on one part of the boundary Γ_s, preventing the structure from undergoing rigid body motion.

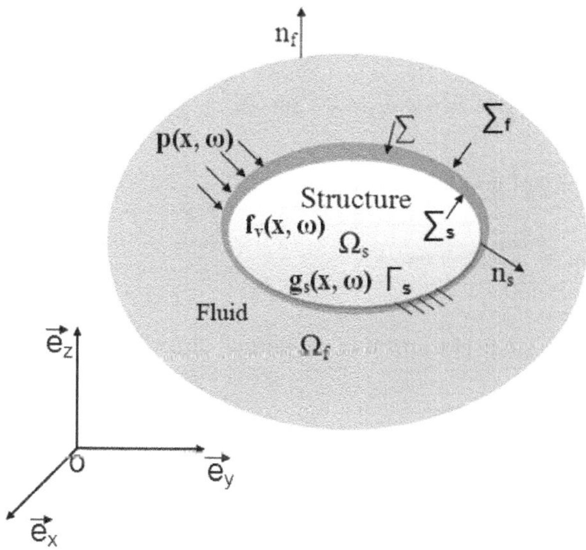

Figure 3.1. *Diagram of the fluid–structure interaction. For a color version of this figure, see www.iste.co.uk/elhami/interactions.zip*

In order to find the discretized equations of the coupled vibroacoustic system, we must first model each of the two components of these system, i.e. the structure and the acoustic cavity.

This study is performed in the frequency domain over the frequency range $[\omega_{\min}; \omega_{\max}]$, where $0 < \omega_{\min} < \omega_{\max}$. The vibroacoustic problem is then reduced to simultaneously solving two problems:

– one for the structure, which is subject to a pressure field p imposed by the fluid on Σ;

– the other for the fluid, which is subject to a displacement field u imposed by the wall Σ.

3.3.2. *Boundary conditions at the interface*

At the interface Σ between the fluid and the structure, the fluid is not viscous and therefore does not adhere to the wall, and so the normal stresses and normal velocities at the wall are continuous, and the tangential stress component is zero (no viscous friction). These boundary conditions may be written as:

– Continuity of normal stresses:

$$\sigma_{ij}.n_i = T_{ij}.n_i = -\rho.\delta_{ij}.n_i, \quad \text{on } \Sigma \tag{3.16}$$

– Continuity of normal velocities:

$$\frac{\partial u_f}{\partial t}.n = \frac{\partial u}{\partial t}.n, \quad \text{on } \Sigma \tag{3.17}$$

– Continuity of displacement in the normal direction:

$$u_f.n = u.n, \quad \text{on } \Sigma \tag{3.18}$$

If we impose a normal acceleration on the interface Σ, we must add an additional condition:

$$\frac{\partial p}{\partial n} = \rho_f.\gamma_n \tag{3.19}$$

which implies that:

$$\frac{\partial p}{\partial n} - \rho_f.\omega^2.un = 0 \quad \text{on } \Sigma \tag{3.20}$$

$$u = 0 \quad \text{on } \Gamma_s \tag{3.21}$$

3.3.3. *Finite element approximation*

3.3.3.1. *Formulation of the non-symmetric coupled fluid–structure problem*

The objective is to determine the equations of the vibroacoustic problem of the structure coupled with its external acoustic cavity. For convenience, we assume that the meshes of the structure and the acoustic cavity are compatible at the interface Σ. If the meshes are incompatible, Guérich and Hamdi [GUE 99] give a way of finding the discretization of the sesquilinear coupling form.

The coupled acoustic fluid–elastic structure problem is described by the following equations: taking the variable p to describe the pressure field in the fluid and u the displacement field in the structure, we have:

$$\frac{\partial \sigma_{ij}(u)}{\partial x_i} + \rho_s.\omega^2.u_i = 0 \quad \text{in} \quad \Omega_s \qquad [3.22]$$

$$\frac{\partial^2 p}{\partial x_i^2} + \frac{\omega^2}{c^2}.p = 0 \quad \text{in} \quad \Omega_f \qquad [3.23]$$

3.3.3.2. *Modeling the structure problem*

To solve the structure problem, we must find a weak solution that satisfies an integral formulation (or variational formulation) equivalent to the initial problem. The weak formulation of the problem is obtained using the method of test functions [BAT 82, SOU 09].

Consider an arbitrary and regular displacement vector v in the domain Ω_s, the virtual displacement field, satisfying the condition $v \;|_{\Gamma_s} = \; 0$. After integrating over the domain Ω_s, weighting the equation of motion of the structure [3.16] by v leads to the following equation:

$$-\omega^2.\int_{\Omega_s} \rho_s.u_i.v_i.dV - \int_{\Omega_s} \frac{\partial \sigma_{ij}(u)}{\partial x_j}.v_i.dV = 0, \qquad [3.24]$$

and we can now apply the formula for generalized integration by parts as follows:

$$\int_{\Omega_s} \frac{\partial \sigma_{ij}(u)}{\partial x_j}.v_i.dV = -\int_{\Omega_s} \sigma_{ij}(u).\frac{\partial}{\partial x_j}(v_i).dV + \int_{\partial\Omega_s} \sigma_{ij}(u).n_j.v_i.dS, \qquad [3.25]$$

From equation [3.16], we can write:

$$\int_{\Omega_s} \frac{\partial \sigma_{ij}(u)}{\partial x_j}.v_i.dV = -\int_{\Omega_s} \sigma_{ij}(u).\varepsilon_{ij}(v).dV$$

$$+\int_{\partial\Omega_s} \sigma_{ij}(u).n_j.v_i.dS. \qquad [3.26]$$

Using the fact that the stress and strain tensors are symmetric, and taking into account the boundary conditions, the variational formulation of the structure problem then becomes:

$$-\omega^2.\int_{\Omega_s} \rho_s.u_i.v_i.dV + \int_{\Omega_s} \sigma_{ij}(u).\varepsilon_{ij}(v).dV = \int_{\partial\Omega_s} \sigma_{ij}(u).n_j.v_i.dS \qquad [3.27]$$

The structure functional may then be written as:

$$F(u) = \frac{1}{2}\int_{\Omega_s} (\sigma_{ij}(u).\varepsilon_{ij}(u) - \rho_s.\omega^2.\langle u,u\rangle)dV - \int_{\partial\Omega_s} p.u.dS. \qquad [3.28]$$

The finite element method can now be applied as usual to solve the variational formulation [3.27].

We wish to find an approximate discrete solution for the continuous problem by working in a discrete geometric domain on which the continuous integral equations can be approximated by algebraic equations. The domain Ω_s is thus approximated by E_s polygonally shaped elementary volumes Ω_{s_e} in such a way that:

$$\Omega_s \sim \cup_{e=1}^{e=E_s} \Omega_{S_e}. \qquad [3.29]$$

This yields a mesh of the structure, and the integrals in formula [3.27] can be approximated as follows:

$$\int_{\Omega_s} \rho_s.u_i.v_i.dV \sim \sum_{e=1}^{e=E_s} \int_{\Omega_{se}} \rho_s.u_i.v_i.dV \qquad [3.30]$$

$$\int_{\Omega_s} \sigma_{ij}(u).\varepsilon_{ij}(v).dV \sim \sum_{e=1}^{e=E_s} \int_{\Omega_s} \sigma_{ij}(u).\varepsilon_{ij}(v).dV_e \qquad [3.31]$$

$$\int_{\partial\Omega_s} p.v_p.dS \sim \sum_{e=1}^{e=E_s} \int_{\partial\Omega_s} p.v_p.dS \qquad [3.32]$$

Now we can calculate the following mass and stiffness terms for each finite element Ω_{s_e}: $\int_{\Omega_{s_e}} \rho_s.u_i.v_i.dV_e$ and $\int_{\Omega_{s_e}} \sigma_{ij}(u).\varepsilon_{ij}(v).dV_e$. The displacement field $u\mid_{\partial\Omega_{s_e}}$ on each finite element Ω_{s_e} is calculated from a finite number of values at given points (nodes) on the finite element using polynomial interpolation. Wc, therefore, obtain a linear equation of the following type on the element Ω_{s_e}:

$$U\mid_{\partial\Omega_{s_e}}(x,t) = N_s^e(x).U_e(t), \qquad [3.33]$$

where U_e is the vector of nodal unknowns for the displacement on the finite element Ω_{s_e} and N_s^e is the matrix of shape functions (interpolation polynomials) for Ω_{s_e}.

Formally, the local unknowns U_e on the finite element Ω_{s_e} can be found from the vector of global unknowns of the problem U by using the matrix equation:

$$U_e = A_s^e.U \qquad [3.34]$$

where A_s^e is the localization matrix indexing the degrees of freedom of the finite element Ω_{s_e} out of the set of degrees of freedom of the problem. We use a similar equation for the virtual fields, namely $V\mid_{\partial\Omega_{s_e}} = N_s^e.V_e$, and then we can calculate the mass and stiffness terms as follows:

3.3.3.2.1. Mass term computation

We approximate the scalar product $u_i.v_i$ on a finite element of the structure, and then we calculate it using the interpolations of the u and v fields:

$$u_i.v_i = V_e^T.N_s^{eT}.N_s^e.U_e, \qquad [3.35]$$

from which we deduce the mass term:

$$\int_{\Omega_{s_e}} \rho_s.u_i.v_i.dV_e = V_e^T\left(\int_{\Omega_{s_e}} \rho_s.N_s^{eT}.N_s^e.dV_e\right).U_e \qquad [3.36]$$

We set: $[M_s^e] = \int_{\Omega_{s_e}} \rho_s.N_s^{eT}.N_s^e.dV_e$ the elementary mass matrix. We thus obtain

$$\int_{\Omega_{s_e}} \rho_s.u_i.v_i.dV_e = V_e^T.[M_s^e].U_e \qquad [3.37]$$

3.3.3.2.2. Stiffness term computation

We approximate the scalar product $\sigma_{ij}(u).\varepsilon_{ij}(v)$ on the finite element Ω_{s_e}. Without loss of generality, we can choose a Cartesian coordinate system. The above scalar product can be written as:

$$\sigma_{ij}.\varepsilon_{ij} = \varepsilon^T.\sigma, \qquad [3.38]$$

where ε are σ are tensors with six components given by:

$$\sigma^T = <\sigma_{xx}, \sigma_{yy}, \sigma_{zz}, \sigma_{xy}, \sigma_{xz}, \sigma_{yz}> \qquad [3.39]$$

$$\varepsilon^T = <\varepsilon_{xx}, \varepsilon_{yy}, \sigma_{zz}, 2\varepsilon_{xy}, 2\varepsilon_{xz}, 2\varepsilon_{yz}> \qquad [3.40]$$

Since the $\sigma(u)$ and $\varepsilon(u)$ tensors are symmetric, equation [3.16] can be written as:

$$\varepsilon = H.U_e, \qquad [3.41]$$

where H is the matrix:

$$H = \begin{pmatrix} 1\ 0\ 0 \\ 0\ 1\ 0 \\ 0\ 0\ 1 \\ 1\ 1\ 0 \\ 1\ 0\ 1 \\ 0\ 1\ 1 \end{pmatrix} .G \qquad [3.42]$$

and

$$G = \begin{pmatrix} \frac{\partial}{\partial x} \\ \frac{\partial}{\partial y} \\ \frac{\partial}{\partial z} \end{pmatrix} \qquad [3.43]$$

is the gradient operator.

Thus, we have: $\varepsilon^e = B^e_s.U_e$ and $B^e_s = H.N^e_s$, where B^e is the matrix of derivatives of the interpolation functions.

From the constitutive law (Hooke's law) for linear elastic bodies, we can write:

$$\sigma = E.\varepsilon \text{ with } E = \begin{pmatrix} L & 0 \\ 0 & M \end{pmatrix} \qquad [3.44]$$

where L and M are given by:

$$L = \begin{pmatrix} \lambda + 2\mu & \lambda & \lambda \\ \lambda & \lambda + 2\mu & \lambda \\ \lambda & \lambda & \lambda + 2\mu \end{pmatrix} \quad \text{and} \quad M = \begin{pmatrix} \mu & 0 & 0 \\ 0 & \mu & 0 \\ 0 & 0 & \mu \end{pmatrix}$$

We can therefore write:

$$\sigma = E.B^e_s.U_e$$

And we can now calculate the stiffness term on a finite element as follows:

$$\begin{aligned} \int_{\Omega_{se}} \sigma_{ij}(u).\varepsilon_{ij}(v).dV_e &= V_e^T \left(\int_{\Omega_{se}} B_s^{eT}.E.B^e_s.dV_e \right) U_e \\ &= V_e^T \left(\int_{\Omega_{se}} N_s^{eT}.H^T.E.H.N^e_s.dV_e \right) .U_e, \end{aligned} \qquad [3.45]$$

We set: $[K^e_s] = \int_{\Omega_{se}} N_s^{eT}.H^T.E.H.N^e_s \, dV_e$ the elementary stiffness matrix. This gives:

$$\int_{\Omega_{se}} \sigma_{ij}(u).\varepsilon_{ij}(v).dV_e = V_e^T [K^e_s] U_e \qquad [3.46]$$

To generalize over the rest of the structure, we simply assemble the elementary stiffness and elementary mass matrices:

$$[K_s] = \sum_e [K_s^e] \qquad [3.47]$$

$$= \sum_e [M_s^e] \qquad [3.48]$$

$$= \sum_e [u^e] \qquad [3.49]$$

The global system to be solved is therefore the following:

$$\left\{[K_s] - \rho_s.\omega^2[M_s]\right\}[u] = [P] \qquad [3.50]$$

3.3.3.3. *For the acoustic cavity*

Let q be an arbitrary but sufficiently regular function in the fluid domain Ω_f. Weighting the acoustic wave equation by q and integrating leads to the following equation:

$$\int_{\Omega_f} (\Delta p + k^2.p).qdV = 0 \qquad [3.51]$$

Integrating this equation by parts gives:

$$\int_{\Omega_f} \left(-\text{grad}(p).\text{grad}(q) + k^2.p.q\right) dV + \int_S q.\frac{\partial p}{\partial n} dS = 0 \text{ with } S = \Sigma \cup \Gamma \qquad [3.52]$$

By setting q to be zero on Γ, and taking into account the boundary conditions on Σ, the variational equation may be written as:

$$\int_{\Omega_f} \left(\text{grad}(p).\text{grad}(q) - k^2.p.q\right) dV = \int_\Sigma \rho_f.c^2.k^2.u_n.q.d\Sigma \qquad [3.53]$$

Note that the left-hand side (B) is a symmetric bilinear form in (p, q) and the right-hand side (L) is a linear form in q. Formally, we can write:

$$B(p, q) = L(q), \quad \forall \text{ statically admissible } q \qquad [3.54]$$

From the symmetry of the bilinear form ($B(p,q) = B(q,p)$), the solution p of the variational equation is a stationary point of the functional:

$$F(p) = \frac{1}{2}B(p,q) - L(p) \qquad [3.55]$$

$$\delta F(p) = \frac{1}{2}(B(p,\delta p) + B(\delta p, p)) - L(\delta p) \qquad [3.56]$$

The stationarity condition of $F(p)$ implies that: $\delta F(p) = 0$, which gives: $B(p,\delta p) = L(\delta p)$. This relation is equivalent to the previous relation, with q playing the role of δp, which is also identically zero on Γ since p is determined on Γ, and we have:

$$B(p,q) = 4.\rho_f.\omega^2.(H(p,q) - Q(p,q)) \qquad [3.57]$$

with

$$H(p,q) = \frac{1}{4.\rho_f.\omega^2}\int_{V_f}(\nabla p.\nabla(q))dV \qquad [3.58]$$

and

$$Q(p,q) = \frac{1}{4}\int_{V_f}\frac{p.q}{\rho_f.c^2}dV \qquad [3.59]$$

The functional of the fluid problem is:

$$F(p) = \frac{1}{2}\int_{V_f}\left(\nabla p.\nabla p - k^2.p.p\right)dV - \int_{\Sigma}\rho_f.c^2.k^2.u_n.p.d\Sigma \qquad [3.60]$$

where $p = p_\Gamma$ on Γ, i.e. $\delta p = 0$ on Γ.

3.3.3.3.1. Discretization by finite elements

Discretizing the variational formulation of the fluid by finite elements gives:

$$H(p,q) = \frac{1}{4\rho_f.\omega^2}\{q\}^T.[H].\{p\} \qquad [3.61]$$

$$Q(p,q) = \frac{1}{4\rho_f.c^2}\{q\}^T.[Q].\{p\} \qquad [3.62]$$

$$L(q) = \rho_f.c^2.k^2\{q\}^T.\{\ddot{u}_n\} \qquad [3.63]$$

$[H]$ and $[Q]$ are the global matrices of the fluid, $\{p\}$ is the nodal pressure of the fluid and $\{\ddot{u}_n\}$ is the nodal vector on the right-hand side.

The fluid functional may be written in the form:

$$F(p) = \frac{1}{2}\{p\}^T.\left([H] - k^2.[Q]\right)\{p\} - \rho_f.c^2.k^2\{p\}^T.\{\ddot{u}_n\} \qquad [3.64]$$

The dynamic matrix equation of the fluid is:

$$([H] - k^2[Q])\{p\} = \rho_f.c^2.k^2.\{\ddot{u}_n\} \qquad [3.65]$$

To determine the acoustic modes of a fluid cavity, it is sufficient to solve the following matrix equation:

$$([H] - k^2[Q])\{p\} = \{0\} \qquad [3.66]$$

3.3.3.4. *For the vibroacoustic problem*

3.3.3.4.1. Variational formulation

The variational formulation of the coupled system is the sum of the two variational equations [3.28] and [5.54]. The solution (u,p) is a stationary point of the functional of the coupled problem:

$$\begin{aligned} F(u,p) &= \frac{1}{2}\int_{\Omega_s}\left(\sigma_{ij}(u)\varepsilon_{ij}(u) - \rho_s.\omega^2.(u,u)\right)dV - \int_\Sigma p.u.d\Sigma \\ &-\frac{1}{2\rho_f.\omega^2}.\int_{V_f}\left[\left(\frac{\partial p}{\partial x_i},\frac{\partial p}{\partial x_i}\right) + k^2.p^2\right]dV \end{aligned} \qquad [3.67]$$

The solution (u,p) is obtained from the variational equation $\delta F(u,p) = 0$.

3.3.3.4.2. Finite element discretization

The surface integral appearing in the formulation [3.67] can be approximated as follows:

$$\int_{\Sigma} p.v.d\Sigma \sim \sum_{e=1}^{e=E_e} \int_{\Sigma_e} p_e.v_e.d\Sigma_e \qquad [3.68]$$

We can write:

$$\int_{\Sigma_e} p_e.v_e.d\Sigma_e = Q_e^T.\left(\int_{\Sigma_e} N_s^{eT}.N_s^e.d\Sigma_e\right).P_e \qquad [3.69]$$

$$= Q_e^T.[C^e].P_e \qquad [3.70]$$

And, summing over the whole surface Σ, we obtain:

$$\int_{\Sigma} p.v.d\Sigma = \{Q^T\}.[C].\{P\} \qquad [3.71]$$

where $[C]$ is the coupling matrix.

Discretizing the weak formulation [3.67] of the coupled problems by finite elements using the operators M_s, K_s, M_f, K_f and C leads to the following algebraic system:

$$\begin{pmatrix} [K_s] - \omega^2[M_s] & -[C] \\ -[C]^T & \frac{1}{\rho_f.\omega^2}.[M_f] - \frac{1}{\rho_f.c^2}.[K_f] \end{pmatrix} \begin{pmatrix} u \\ p \end{pmatrix} = \begin{pmatrix} 0 \\ 0 \end{pmatrix}$$

where:

- $[K_s]$ and $[M_s]$ are the stiffness and mass matrices of the structure;
- $[K_f]$ and $[M_f]$ are the global matrices of the fluid;
- $[C]$ is the coupling matrix;
- u is the nodal displacement vector of the structure;
- p is the nodal pressure vector in the fluid cavity.

3.3.3.5. *Formulation of the symmetric coupled fluid–structure problem*

To symmetrize the system obtained for the coupled problem after discretizing with finite elements, we introduce an additional variable for the fluid (and therefore additional degrees of freedom for the discrete problem), the velocity potential or the displacement potential, denoted as ϕ:

3.3.3.5.1. The velocity potential

The velocity potential satisfies the relation:

$$v = \nabla\phi \qquad \text{where } v \text{ is the velocity,} \tag{3.72}$$

from which we deduce a variational formulation in terms of (u, p, ϕ) that has the useful property of resulting in a symmetric, though large, matrix system after finite element discretization.

Equations of the problem

1) *For the fluid*

We consider small adiabatic perturbations of a perfect fluid around its rest state. The acoustic velocity potential is defined by equation [3.72]. The pressure and potential are related by the following Bernoulli equation:

$$p = -\rho_f.\frac{\partial\phi}{\partial t} \tag{3.73}$$

Moreover, the acoustic equation of the fluid is:

$$\nabla\left(\frac{1}{\rho_f}.\nabla p\right) - \frac{1}{\rho_f.c^2}.\ddot{p} = 0 \tag{3.74}$$

If ρ_f is constant, we find

$$\Delta p - \frac{1}{c^2}\ddot{p} = 0 \tag{3.75}$$

From the two equations [3.73] and [3.74], we deduce:

$$\frac{1}{\rho_f}.\Delta\phi - \frac{1}{\rho_f c^2}\ddot{\phi} = 0 \qquad \text{in } \Omega_f \tag{3.76}$$

This is the wave equation satisfied by the velocity potential.

The boundary conditions of the fluid domain are:

$$\phi = \bar{\phi} \text{ and } \frac{\partial \phi}{\partial n} = \bar{V} \qquad \text{on } \Sigma \qquad [3.77]$$

We write:

- v: the acoustic velocity.
- p: the acoustic pressure.
- ρ_f: the density.
- c : the speed of sound in the domain V_f.
- ϕ: the acoustic velocity potential.

2) *For the structure*

We assume that the structure is linear isotropic, with no initial stress or strain and without any sources (no excitations). Thus, the equation determining its vibrational behavior is given by:

$$\nabla\sigma - \rho_s.\ddot{u} = 0 \qquad \text{in } \Omega_s \qquad [3.78]$$

We write:

- ρ_s: the density of the structure.
- σ: the stress tensor of the structure.
- u: the displacement field of the structure.

The boundary conditions associated with the structure may be stated as:

$$u = u_0 \qquad \text{and} \qquad \sigma.n = p \qquad \text{on } \Sigma \qquad [3.79]$$

3) *Coupling conditions at the fluid–structure interface*

If the fluid and the structure are in contact with each other at the surface Σ, then the coupling conditions on the interface Σ may be written as follows, corresponding to continuity of the normal velocity and continuity of the normal component of the stress tensor:

$$\frac{\partial \phi}{\partial n} = \dot{u}.n \quad \text{and} \quad \sigma.n = \dot{\phi}.n \text{ on } \Sigma \qquad [3.80]$$

4) *Variational formulation of the fluid*

Let ψ be any sufficiently regular test field on the domain Ω_f. Integrating equation [3.76] and multiplying by ψ, we obtain:

$$\int_{V_f} \frac{1}{\rho_f}.\Delta\phi.\psi dV - \int_{V_f} \frac{1}{\rho_f.c^2}.\ddot{\phi}.\psi dV = 0 \quad \forall\psi \ (c.a) \qquad [3.81]$$

Applying the formula for generalized integration by parts:

$$\int_{V_f} \frac{1}{\rho_f}.\Delta\phi.\psi dV = -\int_{V_f} \frac{1}{\rho_f}.\frac{\partial\phi}{\partial x_i}.\frac{\partial\psi}{\partial x_i} dV + \int_{\Sigma} \frac{1}{\rho_f}.\frac{\partial\phi}{\partial x_i}.n_i^f.\psi dS \qquad [3.82]$$

This gives:

$$-\int_{V_f} \frac{1}{\rho_f}.\frac{\partial\phi}{\partial x_i}.\frac{\partial\psi}{\partial x_i} dV + \int_{\Sigma} \frac{1}{\rho_f}.\frac{\partial\phi}{\partial x_i}.n_i^f.\psi dS - \int_{V_f} \frac{1}{\rho_f.c^2}.\ddot{\phi}.\psi dV = 0 \qquad [3.83]$$

Taking into account the boundary conditions [3.77] and [3.80], equation [3.83] becomes:

$$\int_{V_f} \frac{1}{\rho_f}.\frac{\partial\phi}{\partial x_i}.\frac{\partial\psi}{\partial x_i} dV + \int_{V_f} \frac{1}{\rho_f.c^2}.\ddot{\phi}.\psi dV - \int_{\Sigma} \dot{u}^s.n^f.\psi dS = \int_{\Sigma} \dot{u}^s.n^f.\psi dS - \int_{\Omega_f} \bar{V}.\psi dV \quad \forall\psi \ c.a. \qquad [3.84]$$

5) *Variational formulation of the structure*

Let v be any sufficiently regular test field on the domain Ω_s. Integrating equation [3.78] and multiplying by v, we find:

$$\int_{\Omega_s} \nabla\sigma.v.dV - \int_{\Omega_s} \rho_s.\ddot{u}.v.dV = 0 \text{ on } \Omega_s \qquad [3.85]$$

[3.86]

Hence:

$$\int_{\Omega_s} \frac{\partial\sigma_{ij}(u_i)}{\partial x_i}.v_i.dV - \int_{\Omega_s} \rho_s.\ddot{u}_i.v_i.dV = 0 \qquad [3.87]$$

From the formula for generalized integration by parts, we deduce that:

$$\int_{\Omega_s} \frac{\partial \sigma_{ij}(u_i)}{\partial x_i}.v_i.dV = -\int_{\Omega_s} \sigma_{ij}(u_i)\varepsilon_{ij}(v_i)dV + \int_{\Sigma} \sigma_{ij}(u_i)n_j^s.v_i dS \quad [3.88]$$

Substituting into equation [3.87], and taking into account the boundary conditions [3.79], we find:

$$\int_{\Omega_s} \rho_s.\ddot{u}_i.v_i.dV + \int_{\Omega_s} \sigma_{ij}(u_i)\varepsilon_{ij}(v_i)dV - \int_{\Sigma} \sigma_{ij}(u_i).n_j^s.v_i dS = \int_S v.\bar{p}dS \quad [3.89]$$

Given the boundary conditions [3.80], equation [3.89] becomes:

$$\int_{\Omega_s} \rho_s.\ddot{u}.v.dV + \int_{\Omega_s} \sigma(u)\varepsilon(v)dV - \int_{\Sigma} \dot{\phi}.n.vdS = \int_S v.\bar{p}dS \quad [3.90]$$

6) *Discretization by finite elements*

Discretizing the mass and stiffness terms of the fluid and structure problems using the finite element method allows us to calculate the matrices M_f, M_s, K_f, K_s. The formulation [1.38] contains additional terms that can be discretized using the operators defined earlier:

$$\int_{V_f} \frac{1}{\rho_f}.\ddot{\phi}.\psi dV \hookrightarrow \psi^T.[M_f].\left\{\ddot{\Phi}\right\} \quad [3.91]$$

$$\int_{V_f} \frac{1}{\rho_f}.\frac{\partial \phi}{\partial x_i}.\frac{\partial \psi}{\partial x_i}dV \hookrightarrow \psi^T.[K_f].\{\Phi\} \quad [3.92]$$

Discretizing the term describing the action of the structure on the fluid gives:

$$\int_{\Sigma} \dot{u}.n^f.\psi dS \hookrightarrow \psi^T.[C]^T.\left\{\dot{U}\right\} \quad [3.93]$$

$$\int_{\Sigma} \bar{v}.\psi.dS \hookrightarrow \psi^T.\{V^f\} \quad [3.94]$$

Discretizing the formulation [1.38] gives:

$$\psi^T.\left\{[M_f].\left\{\ddot{\Phi}\right\} + [K_f].\{\Phi\} + [C].\left\{\dot{U}\right\}\right\} = \psi^T.\{V^f\} \quad [3.95]$$

The formulation [3.90] contains additional terms that may be discretized as follows, using the operators defined earlier:

$$\int_{\Omega_s} \rho_s.\ddot{u}.vdV \hookrightarrow V^T.[M_s].\left\{\ddot{U}\right\} \qquad [3.96]$$

$$\int_{\Omega_s} \sigma(u).\varepsilon(v)dV \hookrightarrow V^T.[K_s].\{U\} \qquad [3.97]$$

Discretizing the term describing the action of the fluid on the structure gives:

$$\int_{\Sigma} \dot{\phi}.n.vdS \hookrightarrow V^T.[C].\left\{\dot{\Phi}\right\} \qquad [3.98]$$

$$\int_{S} V.\bar{P}.dS \hookrightarrow V^T.\{P_s\} \qquad [3.99]$$

Discretizing the formulation [3.90] gives:

$$V^T.\left\{[M_s].\left\{\ddot{U}\right\} + [K_s].\{U\} + [C].\left\{\dot{\Phi}\right\}\right\} = V^T.\{P\} \qquad [3.100]$$

We assemble the structure domain and the acoustic domain into a global vector with all of the degrees of freedom of both the fluid and the structure, organized as follows:

$$U = (u, \phi) \qquad [3.101]$$

Now, noting that the test vector (v, ψ) is arbitrary, we can show that these assembled formulations [3.95] and [3.100] lead to the following algebraic system:

$$[M].\left\{\ddot{U}\right\} + [C].\left\{\dot{U}\right\} + [K].\{U\} = \{f\} \qquad [3.102]$$

where

$$[M] = \begin{pmatrix} [M_s] & 0 \\ 0 & [M_f] \end{pmatrix}, \qquad [K] = \begin{pmatrix} [M_s] & 0 \\ 0 & [M_f] \end{pmatrix} \qquad [3.103]$$

$$[C] = \begin{pmatrix} 0 & C \\ T & 0 \end{pmatrix}, \{f\} = \left\{ \begin{matrix} P \\ V_f \end{matrix} \right\} \quad [3.104]$$

3.3.3.5.2. The displacement potential

Keeping the same objective as above, namely to symmetrize the system obtained by finite element discretization, a symmetric form of the elasto-acoustic problem can be found by using a coupled formulation that represents the structure problem by the displacement u and the fluid problem by the pressure p and the displacement potential ϕ. This is given by the relation:

$$u = \nabla \phi \quad [3.105]$$

Equations of the problem

The equations of the coupled fluid–structure problem are given by:

1) *For the fluid*

The displacement potential ϕ is related to the pressure by:

$$p = -\rho_f.\omega^2.\phi \quad [3.106]$$

The acoustic equation of the fluid is:

$$\Delta p + k^2.p = 0 \qquad \text{in } \Omega_f \text{ with } k^2 = \frac{\omega^2}{c^2} \quad [3.107]$$

From two equations [3.106] and [3.107], we can write:

$$\Delta \left(\rho_f.\omega^2.\phi\right) + \frac{\omega^2}{c^2}.p = 0 \quad [3.108]$$

[3.109]

If ρ_f is constant, we have that:

$$\Delta\phi + \frac{p}{\rho_f.c^2} = 0 \quad [3.110]$$

Hence

$$\frac{p}{\rho_f.c^2} - \frac{\omega^2}{c^2}.\phi = 0 \qquad \text{in } \Omega_f$$

therefore the fluid equations are:

$$p - \rho_f.\omega^2.\phi = 0 \tag{3.111}$$

$$\Delta\phi + \frac{p}{\rho_f.c^2} = 0 \tag{3.112}$$

2) *For the structure*

The equation of motion of the structure is:

$$\frac{\partial\sigma_{ij}(u)}{\partial x_j} + \rho_s.\omega^2.u_i = 0 \quad \text{in } \Omega_s \tag{3.113}$$

Assuming that the structure has loads applied to its outer surface, we have:

$$\sigma_{ij}.n_j + \bar{p}.n_i = 0. \tag{3.114}$$

3) *Coupling conditions*

The fluid–structure coupling conditions are:

$$\sigma_{ij}.n_j^s = \rho_f.\omega^2.\phi.n_i \qquad \text{on } \Sigma, \tag{3.115}$$

$$\frac{\partial\phi}{\partial x_i}.n_j = u_j.n_j \qquad \text{on } \Sigma, \tag{3.116}$$

hence:

$$\frac{\partial\phi}{\partial n} = u_n^s \quad \text{on } \Sigma, \tag{3.117}$$

These conditions state that the normal component of the stress tensor and the normal component of the displacement are continuous at the fluid–structure interface.

4) *Variational formulation of the fluid*

The variational formulation of the fluid is obtained by multiplying equation [3.111] by a virtual pressure field q and a virtual displacement potential field ψ.

Integrating by parts on the fluid domain V_f and taking into account the terms corresponding to the boundary conditions, we obtain the following integral forms:

$$\frac{1}{\rho_f.c^2}\left\{\int_{\Omega_f} p.qdV - \int_{\Omega_f} \rho_f.\omega^2.\phi.qdV\right\} = 0, \qquad [3.118]$$

hence:

$$\frac{1}{\rho_f.c^2}\int_{V_f} \left(p - \rho_f.\omega^2.\phi\right).qdV = 0 \;\forall \text{ statically admissible } q \qquad [3.119]$$

and

$$\rho_f \int_{\Omega_f} \frac{\partial\phi}{\partial x_i}.\frac{\partial\psi}{\partial x_i}dV - \int_{\Omega_f} \frac{p.\psi}{c^2}dV - \rho_f \int_{\Sigma} u_n.\psi d\Sigma = 0 \;\forall\psi \qquad [3.120]$$

5) *Variational formulation of the structure*

The variational formulation of the structure problem is obtained by multiplying equation [3.113] by a virtual displacement field v. Integrating by parts on the domain Ω_s, and taking into account the boundary conditions [3.114] and [3.117], we obtain the following integral form:

$$\int_{\Omega_s} \sigma_{ij}(u)\varepsilon_{ij}(v) - \rho_s.\omega^2.u.vdV = \int_{\Sigma} \rho_f.\omega^2.\phi.vd\Sigma + \int_S \rho_f.\bar{p}.vdS \qquad [3.121]$$

6) *Variational formulation of the coupled problem*

By summing three equations [3.119]–[3.121], we obtain the following variational equality:

$$\mathcal{L}\left((u,p,\phi),(v,q,\psi)\right) = L(v), \quad \forall\,(v,q,\psi), \qquad [3.122]$$

where $\mathcal{L}$ is the symmetric bilinear form given by:

$$\mathcal{L}\left((u,p,\phi),(v,q,\psi)\right) = \int_{V_f}\left(\rho_f.\omega^2\frac{\partial\phi}{\partial x_i}.\frac{\partial\psi}{\partial x_i} - k^2(\phi.q + p.\psi) + \frac{1}{\rho_f.c^2}.p.q\right)dV$$

$$+\int_{\Omega_s} \sigma_{ij}(u)\varepsilon_{ij}(v) - \rho_s.\omega^2.(u,v)dV - \rho_f.\omega^2 \int_\Sigma (\phi v + u\psi)d\Sigma \quad [3.123]$$

and L is the linear form associated with the external load, which is assumed to be known:

$$L(v) = \int_S \bar{p}.vdS \quad [3.124]$$

The functional of the coupled fluid–structure problem is:

$$F(u,p,\phi) = \frac{1}{2}\int_{V_f}\left(\rho_f.\omega^2\frac{\partial\phi}{\partial x_i}.\frac{\partial\phi}{\partial x_i} - 2k^2\phi.p + \frac{1}{\rho_f.c^2}.p.p\right)dV$$
$$+\frac{1}{2}.\int_{\Omega_s}\sigma_{ij}(u)\varepsilon_{ij}(v) - \rho_s.\omega^2.(u.u)dV - 2\rho_f.\omega^2\int_\Sigma \phi ud\Sigma - \int_S \bar{p}.udS \quad [3.125]$$

(u,p,ϕ) is a stationary point of the function, i.e. $\delta F(u,p,\phi) = 0$.

7) *Discretization by finite elements*

Discretizing the mass and stiffness terms of the fluid and structure problems using the finite element method allows us to calculate the matrices M_s, M_f, K_s and K_f. The formulation [3.125] contains additional terms that may be discretized using the operators defined earlier:

$$\int_{V_f}\nabla\phi.\nabla\psi.\psi dV \hookrightarrow \Psi^T.[K_f].\Phi \quad [3.126]$$

$$\int_{V_f}\frac{p.\psi}{c^2}dV \hookrightarrow \Psi^T.[M_f].P \quad [3.127]$$

$$\int_{V_f}\frac{\phi.q}{c^2}dV \hookrightarrow Q^T.[M_f].\Phi \quad [3.128]$$

$$\int_\Sigma \phi.n_i.q_i dS \hookrightarrow Q^T.[C].\Phi \quad [3.129]$$

$$\int_\Sigma u_i.n_i.\psi dS \hookrightarrow \Psi^T.[C]^T.U \quad [3.130]$$

$$\int_{\Omega_s}\sigma_{ij}(u).\varepsilon_{ij}(v)dV \hookrightarrow V^T.[M_s].U \quad [3.131]$$

$$\int_{\Omega_s} u.vdV \hookrightarrow V^T.[K_s].U \qquad [3.132]$$

$$\int_S v.\bar{p}.dS \hookrightarrow V^T.\{f\} \qquad [3.133]$$

From the above, we deduce the following equations:

$$-\omega^2.V^T.[M_s].U + V^T.[K_s].U + \rho_F.V^T.[C].\Phi = 0 \qquad [3.134]$$

$$\frac{1}{\rho_f}.Q^T.[M_f].P - \omega^2.\Psi^T.[M_f].\Phi = 0 \qquad [3.135]$$

$$-\rho_f.\Psi^T.[K_f].\Phi + \Psi^T.[M_f].P + \rho_f.\Psi^T.[C]^T.U = 0 \qquad [3.136]$$

These equations hold for all v, q and ψ.

Multiplying the last equation by $(-\omega^2)$, we obtain the following symmetric matrix system:

$$\begin{pmatrix} [K_s] & 0 & 0 \\ 0 & \frac{1}{\rho_f}.[M_f] & 0 \\ 0 & 0 & 0 \end{pmatrix} . \begin{Bmatrix} U \\ P \\ \Phi \end{Bmatrix} - \omega^2 \begin{pmatrix} [M_s] & 0 & \rho_f.[C] \\ 0 & 0 & [M_f] \\ \rho_f.[C]^T & [M_f] & -\rho_f.[K_f] \end{pmatrix} . \begin{Bmatrix} U \\ P \\ \Phi \end{Bmatrix} = \begin{Bmatrix} F \\ 0 \\ 0 \end{Bmatrix} \qquad [3.137]$$

If there is no load, system [3.137] reduces to:

$$\begin{pmatrix} [K_s] & 0 & 0 \\ 0 & \frac{1}{\rho_f}.[M_f] & 0 \\ 0 & 0 & 0 \end{pmatrix} . \begin{Bmatrix} U \\ P \\ \Phi \end{Bmatrix} - \omega^2 \begin{pmatrix} [M_s] & 0 & \rho_f.[C] \\ 0 & 0 & [M_f] \\ \rho_f.[C]^T & [M_f] & -\rho_f.[K_f] \end{pmatrix} . \begin{Bmatrix} U \\ P \\ \Phi \end{Bmatrix} = \begin{Bmatrix} 0 \\ 0 \\ 0 \end{Bmatrix} \qquad [3.138]$$

From the last row of system [3.138], we deduce that:

$$\Phi = \frac{1}{\rho_f}.[K_f]^{-1}.[M_f].[P] + [K_f]^{-1}.[C]^T.[U] \qquad [3.139]$$

Substituting equation [3.139] into the last two rows of system [3.138], we can eliminate the ϕ variable, so that the system becomes:

$$\begin{pmatrix} [K_s] & 0 \\ 0 & \frac{1}{\rho_f}.[M_f] \end{pmatrix} . \begin{Bmatrix} U \\ P \end{Bmatrix} - \omega^2 \begin{bmatrix} [M_s] + \rho_f.[e].[C]^T & [e].[M_f] \\ .[C]^T & \frac{1}{\rho_f}.[m].[M_f] \end{bmatrix} . \begin{Bmatrix} U \\ P \end{Bmatrix} = \begin{Bmatrix} 0 \\ 0 \end{Bmatrix} \quad [3.140]$$

where $[e] = [C].[K_f]^{-1}$ and $[m] = [M_f].[K_f]^{-1}$.

3.4. Study of an elastic plate coupled with a fluid cavity

This section is dedicated to studying the vibro-acoustic behavior of an elastic plate coupled with a fluid cavity (Figure 3.2).

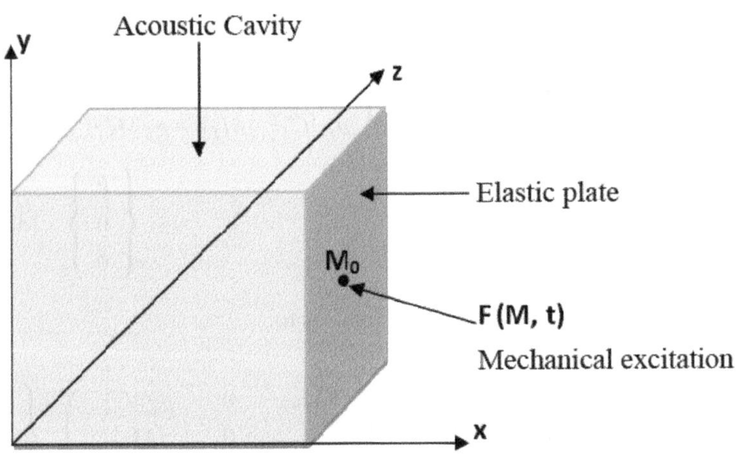

Figure 3.2. *Elastic plate coupled with a fluid cavity*

In this study, we consider a thin plate that is only supported at its four edges. e is the thickness of the plate. Our working hypotheses are the following:

1) we will work within the framework of linear elasticity and linear acoustics;

2) we will consider a steady state (harmonic with respect to time).

In this study, the mechanical excitation is modeled as a point force applied to the elastic plate at the point M_0. This force can be written as:

$$F(M,t) = F_x(M,t) = F_0.\delta.(M - M_0).e^{i\omega t} \qquad [3.141]$$

where $\delta.(M - M_0)$ is the Dirac function.

3.4.1. *Equations of the coupled fluid–structure problem*

The equation of motion of the thin elastic plate is:

$$D.\Delta\Delta w - \sigma\,\omega^2.w \;-\; F_0.\delta.(M - M_0) + p \quad \text{in} \quad (\Sigma) \qquad [3.142]$$

where D is the flexural modulus of the plate, σ is the mass per unit area of the plate, $p(x, y, z)$ is the fluid pressure and $w(y, z)$ is the normal displacement (arrow) of the elastic plate.

The boundary conditions on all edges of the plate are:

$$w = 0 \qquad [3.143]$$

The equation of motion of the fluid cavity is:

$$(\Delta + k^2).p = 0 \quad \text{in} \quad (\Omega_f) \qquad [3.144]$$

where k is the acoustic wavenumber, $k = \frac{\omega}{c}$ and c is the speed of sound in the fluid.

The boundary conditions on all rigid faces of the fluid cavity are as follows:

$$\frac{\partial p}{\partial n} = 0 \qquad [3.145]$$

The fluid–structure coupling conditions at the surface of interaction are:

$$\frac{\partial p}{\partial n} - \rho_f.\omega^2.w = 0 \quad \text{on} \quad \Sigma \qquad [3.146]$$

$$\sigma_{ij}.n_j + p.n_i = 0 \quad \text{on} \quad \Sigma \qquad [3.147]$$

where ρ_f is the density of the fluid and n is the outward normal of the surface Σ.

3.4.2. *Variational formulation of the fluid*

In this formulation, we use a modal presentation to describe the matrix equation of motion of the plate coupled with the fluid cavity in terms of the modal basis of the structure and the acoustic modal basis.

The fluid pressure decomposes as follows with respect to its modal basis:

$$p(x, y, z) = \sum_{n} \alpha_n.q_n = \sum_{ijl} \alpha_{ijl}.q_{ijl} \qquad [3.148]$$

with

$$q_{ijl} = \cos\frac{i\pi}{a}x.\cos\frac{j\pi}{b}y.\cos\frac{l\pi}{c}z \qquad [3.149]$$

For any test function q_m, we can write:

$$\int_{\Omega_f} q_m.(\Delta + k^2).pd\Omega_f \quad \text{in} \quad (\Omega_f) \qquad [3.150]$$

q_m is the generalized vector corresponding to the acoustic mode m.

Integrating this equation by parts gives:

$$\int_{\Omega_f} (-(\text{grad}(q_m), \text{grad}(p)) + k^2 q_m.p)d\Omega_f + \int_{\partial\Omega_f} q_m.\text{grad}(p).nd\partial\Omega_f = 0 \qquad [3.151]$$

$\partial\Omega_f$ corresponds to the boundaries of the fluid cavity Ω_f.

Applying the boundary conditions and fluid–structure coupling conditions, we have:

$$\int_{\Omega_f} (-(\mathrm{grad}(q_m), \mathrm{grad}(p)) + k^2 q_m.p) d\Omega_f + \rho_f.\omega^2 \int_{\Sigma} q_m.w d\Sigma = 0 \quad [3.152]$$

The normal displacement of the elastic plate decomposes as follows with respect to its modal basis:

$$w(y, z) = \sum_r \beta_r.W_r = \sum_{pq} \beta_{pq}.W_{pq} \quad [3.153]$$

with

$$W_{pq} = \sin \frac{p\pi}{b} y.sin \frac{q\pi}{c} z \quad [3.154]$$

The variational equation becomes:

$$\int_{\Omega_f} (-(grad(q_m), grad(\Sigma_n \alpha_n.q_n)) + k^2 q_m.\Sigma_n \alpha_n.q_n) d\Omega_f + \rho_f.\omega^2 \int_{\Sigma} q_m.\Sigma_r \beta_r.W_r d\Sigma = 0 \quad [3.155]$$

This equation may be written in the form:

$$\sum_n \alpha_n \int_{\Omega_f} (-(\mathrm{grad}(q_m), \mathrm{grad}(q_n)) + k^2 q_m.q_n) d\Omega_f + \rho_f.\omega^2 \Sigma_r \beta_r \int_{\Sigma} q_m.W_r d\Sigma \quad [3.156]$$

We know that:

$$\int_{\Omega_f} (\mathrm{grad}(q_m), \mathrm{grad}(q_n)) d\Omega_f = \int_{\partial\Omega_f} \mathrm{div}(q_m.\mathrm{grad}(q_n)) d\partial\Omega_f - \int_{\Omega_f} q_m.\Delta q_n d\Omega_f \quad [3.157]$$

From the Helmotz equation, we can write:

$$\Delta q_n = -k_n^2 q_n \qquad [3.158]$$

which implies that

$$\sum_n \alpha_n \int_{\partial\Omega_f} -\mathrm{div}(q_m.\frac{\partial q_n}{\partial n})d\partial\Omega_f + \sum_n \alpha_n \int_{\Omega_f} q_m.\Delta q_n d\Omega_f + \sum_n \alpha_n \int_{\Omega_f} k^2 q_m.q_n d\Omega_f + \rho_f.\omega^2 \sum_r \beta_r \int_\Sigma q_m.W_r d\Sigma = 0 \qquad [3.159]$$

This equation may also be written in the form:

$$\sum_n \alpha_n \int_{\Omega_f} (k^2 - k_n^2) q_m.q_n d\Omega_f + \rho_f.\omega^2 \sum_r \beta_r \int_\Sigma q_m.W_r d\Sigma = 0 \qquad [3.160]$$

with

$$\int_{\Omega_f} q_m.q_n d\Omega_f = \delta_{mn}.\mathrm{cte} \qquad [3.161]$$

cte is a constant that depends on i, j and l, and δ_{mn} is the Kronecker symbol:

$$\delta_{mn} = \begin{cases} 0 & \text{if} \quad m \neq n \\ 1 & \text{if} \quad m = n \end{cases}$$

3.4.2.1. *Calculating the constant:*

cte

$- i = j = l = 0$

$$\int_{\Omega_f} q^2(x, y, z)d\Omega_f = a.b.c = V \qquad [3.162]$$

V is the volume of the fluid cavity.

– $i = j = 0, l \neq 0$ or $i = l = 0, j \neq 0$ or $l = j = 0, i \neq 0$

$$\int_{\Omega_f} q^2(x,y,z)d\Omega_f = \frac{a.b.c}{2} = \frac{V}{2} \qquad [3.163]$$

– $i = 0, j \neq 0, l \neq 0$ or $j = 0, i \neq 0, l \neq 0$ or $l = 0, i \neq 0, j \neq 0$

$$\int_{\Omega_f} q^2(x,y,z)d\Omega_f = \frac{a.b.c}{4} = \frac{V}{4} \qquad [3.164]$$

– $i \neq 0, j \neq 0, l \neq 0$

$$\int_{\Omega_f} q^2(x,y,z)d\Omega_f = \frac{a.b.c}{8} = \frac{V}{8} \qquad [3.165]$$

Equation [3.160] can now once again be written in the form:

$$(k^2 - k_n^2)\alpha_n.Cte + \rho_f.\omega^2\Sigma_r\beta_r.C_{nr} = 0 \qquad [3.166]$$

with

$$C_{nr} = \int_{\Sigma} q_n.W_r d\Sigma \qquad [3.167]$$

3.4.2.2. *Calculating the modal coupling matrix* C_{nr}

For $x = 0$, we have:

$$q_{ijl} = \cos\frac{j\pi}{b}y.\cos\frac{l\pi}{c}z \qquad [3.168]$$

hence,

$$C_{nr} = \int_{\Sigma} \sin\frac{p\pi}{b}y.\cos\frac{j\pi}{b}y.\sin\frac{q\pi}{c}z.\cos\frac{l\pi}{c}zdydz \qquad [3.169]$$

The modal coupling matrix is given by the equation:

$$C_{nr} = \frac{b.c.p.q}{\pi^2.(p^2 - j^2).(q^2 - l^2)} y.[(1 - (-1)^{p+j}.(1 - (-1)^{q+l}))] \qquad [3.170]$$

If $p = j$ or $q = l, \Rightarrow C_{nr} = 0$.

3.4.3. *Variational formulation of the plate*

For any kinematically admissible test function w_m, we can write:

$$\int_{\Sigma} w_m. \left(D\Delta\Delta w - \sigma.\omega^2.w\right) d\Sigma = w_m(X_0).F_0 + \int_{\Sigma} w_m.p.d\Sigma \qquad [3.171]$$

w_m is the generalized displacement vector corresponding to the structural mode m.

From Green's formula, we have:

$$\int_{\Sigma} D.w_m.\Delta\Delta w d\Sigma = -\int_{\Sigma} D.grad(w_m).grad(\Delta w)d\Sigma + \int_{\partial\Sigma} D.w_m.grad(\Delta w)d\partial\Sigma \qquad [3.172]$$

and

$$\int_{\Sigma} D.grad(w_m).grad(\Delta w)d\Sigma = -\int_{\Sigma} D.\Delta w_m.\Delta w d\Sigma + \int_{\partial\Sigma} D.grad(w_m).\Delta w d\partial\Sigma \qquad [3.173]$$

This gives

$$\int_{\Sigma} D.w_m.\Delta\Delta w d\Sigma = \int_{\Sigma} D.\Delta w_m.\Delta w d\Sigma \qquad [3.174]$$

Equation [3.171] becomes:

$$\int_\Sigma D.\Delta w_m.\Delta w d\Sigma - \int_\Sigma w_m.\sigma.\omega^2.w d\Sigma = w_m(X_0).F_0 + \int_\Sigma w_m.p.d\Sigma \quad [3.175]$$

Replacing the displacement vector of the plate and the pressure in the fluid cavity by their respective expressions, we can write

$$\int_\Sigma D.\Delta w_m.\Delta(\sum_r \beta_r.w_r)d\Sigma - \int_\Sigma w_m.\sigma.\omega^2.\sum_r \beta_r.w_r d\Sigma = w_m(X_0).F_0 + \int_\Sigma w_m.\sum_n \alpha_n.q_n.d\Sigma \quad [3.176]$$

For $m = r$, we have

$$\beta_r.\int_\Sigma (D.\Delta w_m.\Delta w_r - \sigma.\omega^2.w_r.w_r)d\Sigma = w_r(X_0).F_0 + \int_\Sigma w_r.\sum_n \alpha_n.q_n.d\Sigma \quad [3.177]$$

Replacing the modal displacement vector of the plate by its expression, we can write

$$\beta_r.\int_\Sigma \left(D.(\frac{p^2\pi^2}{b^2} + \frac{q^2\pi^2}{c^2})w_r.w_r - \sigma.\omega^2.w_r.w_r\right) d\Sigma = w_r(X_0).F_0 + \int_\Sigma w_r.\sum_n \alpha_n.q_n.d\Sigma \quad [3.178]$$

We set:

$$\omega_r^2 = \frac{D}{\sigma}.\left(\frac{p^2\pi^2}{b^2} + \frac{q^2\pi^2}{c^2}\right) \quad [3.179]$$

We then have

$$\beta_r. \int_{\Sigma} \left((\omega_r^2 - \omega^2).w_r.w_r \right) d\Sigma \qquad [3.180]$$

$$= w_r(X_0).\frac{F_0}{\sigma} + \frac{1}{\sigma}.\sum_n \alpha_n. \int_{\Sigma} w_r.q_n.d\Sigma \qquad [3.181]$$

We know that

$$\int_{\Sigma} w_r.w_r d\Sigma = \frac{bc}{4} = \Sigma_p \qquad [3.182]$$

where Σ_p is the surface of the plate.

Finally, we have

$$(\omega_r^2 - \omega^2).\beta_r - \frac{4}{\sigma.\Sigma_p}.\sum_n \alpha_n.C_m = w_r(X_0).\frac{4.F_0}{\sigma.\Sigma_p} \qquad [3.183]$$

Using equations [3.182], the modal equation of motion of the plate coupled with the fluid cavity may be written as:

$$\begin{bmatrix} \omega_r^2 - \omega^2 & \frac{-4}{\sigma.\Sigma_p}.C_m \\ \frac{-4}{\sigma.\Sigma_p}.C_{nr} & (k_n^2 - k^2).\frac{4.Cte}{\rho_f.\omega^2.\sigma.\Sigma_p} \end{bmatrix} \begin{Bmatrix} \beta_r \\ \alpha_n \end{Bmatrix} = \begin{Bmatrix} F_s \\ 0 \end{Bmatrix} \qquad [3.184]$$

with $F_s = w_r(X_0).\frac{4.F_0}{\sigma.\Sigma_p}$.

3.4.4. *Numerical results*

3.4.4.1. *For the 2D plate*

This example illustrates a steady-state problem with vibroacoustic interactions and demonstrates how coupling may be used in situations where the structural deformations are small and linear. The plate is elastic (with Young's modulus $E = 2.11 \times 10^{11}$, density $\rho_s = 7,860\,\text{kg.m}^{-3}$ and pressure coefficient $\nu = 0.3$) and is fully immersed in a compressible fluid at rest (with density $\rho_f = 1,000$ kg.m^{-3} and speed of sound $c = 1,500\,\text{m}^{s-1}$). The objective of the problem is to determine the natural frequencies of the structure coupled with the water, and the effect of the coupling on the vibrations of the system, without appealing to model reduction [MAN 13a].

The structure is assumed to be mounted from the top end and is not subject to a load, so that the fluid may be assumed to be immobile at the edges and the pressure is also zero at the edges, except on the top end, which is assumed to be free. The problem is modeled using commercial Ansys© code.

To compute the finite elements, the fluid and the structure are meshed with rectangular elements of four nodes (FLUID29) and (PLANE42), respectively. The chosen mesh is a compatible mesh, so that the nodes of the two domains coincide at the fluid–structure interface, with 600 elements, 100 for the structure and 500 for the fluid. Each element has four nodes and each node has three degrees of freedom U_x and U_y (displacement in the x and y directions) and P, the pressure at the interface. Figure 3.3 shows the model of the coupled problem and its finite element approximation [MAN 13a].

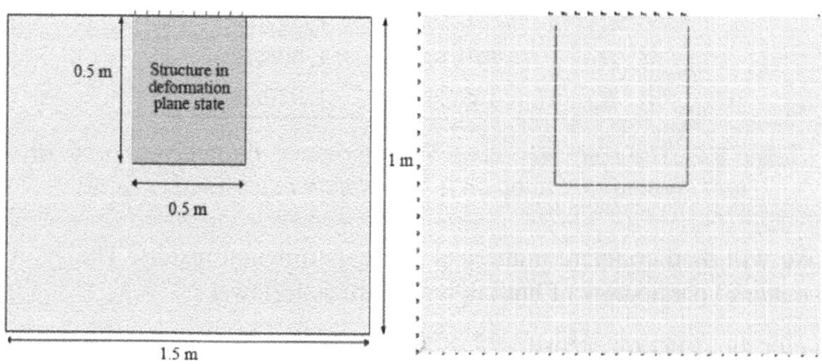

Figure 3.3. *Finite element model of the system. For a color version of this figure, see www.iste.co.uk/elhami/interactions.zip*

The results obtained for the structure, both dry and immersed, are shown in Tables 3.1 and 3.5.

Note that the results found by both methods are in good agreement, but require very high computation times.

3.4.4.2. *For the 3D plate*

This section proposes a deterministic study of a thin 3D structure that may be viewed as a membrane. The structure is mounted at the top end (along its width), and is not subject to any load. This problem illustrates the use of coupling with small linear structural deformations. The plate is thin, with a thickness of 1 mm, and is fully immersed in the water, which is at rest. The objective is to determine the first five natural frequencies.

Modes	Dry structure	Immersed structure
R_1	$1,122$	890.12
R_2	$2,743.9$	998.16
R_3	$3,016$	$1,312.2$
R_4	$4,850.1$	$1,595.7$
R_5	$5,108.6$	$1,930.9$

Table 3.1. *The first five natural frequencies of the structure, both dry and immersed, based on a formulation in terms of* (u,p)

Modes	Dry structure	Immersed structure
R_1	$1,122.1$	890.14
R_2	$2,743.93$	998.15
R_3	$3,016.15$	$1,312.92$
R_4	$4,850.16$	$1,595.73$
R_5	$5,108.66$	$1,930.94$

Table 3.2. *The first five natural frequencies of the structure, both dry and immersed, based on a formulation in terms of* (u,p,Φ)

We will then compare numerical and experimental results. The geometry and material parameters of both domains are as follows:

– for the structure: density = $7,800$ kg.m^{-3}; Young's modulus = 2.1×10^{11} Pa; Poisson coefficient = 0.3; length = 9.6 cm; width = 7.2 cm; height = 1 mm;

– for the fluid: density = $1,000$ kg.m^{-3}; speed of sound = $1,500$ m.s^{-1} ; length = 28.5 cm; width = 14 cm; height = 19 cm.

The chosen mesh is tetrahedral with 800 elements. Each element has four nodes, and each node has six degrees of freedom U_x, U_y and U_z (displacement in the x, y and z directions), rotx, roty and rotz (rotation along x, y and z). Figure 3.4 shows a diagram of the model and its finite element discretization.

The results obtained without model reduction for the structure, both dry and immersed, are shown in Tables 3.3 and 3.4. The experimental results are taken from [BOR 06].

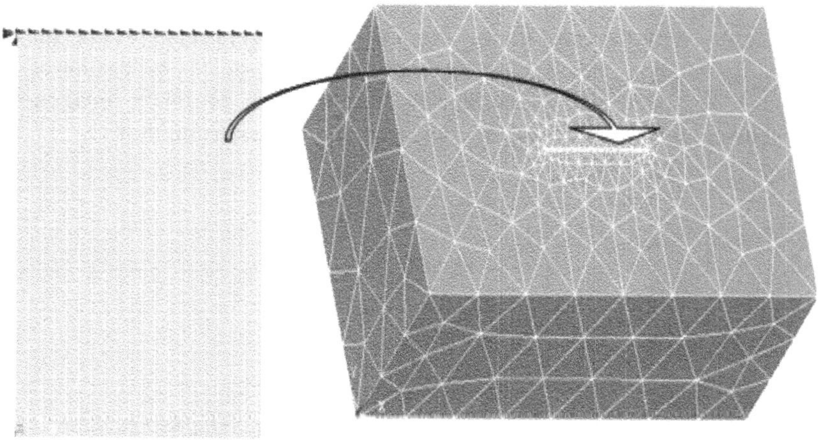

Figure 3.4. *Finite element discretization of tho 3D plate, both dry and immersed. For a color version of this figure, see www.iste.co.uk/elhami/interactions.zip*

	Ansys	Experimental
f_1	90.12	87.9
f_2	279.64	280
f_3	552.38	548
f_4	960.93	–
f_5	1,162.74	1,184

Table 3.3. *The first five natural frequencies of the dry structure*

3.5. Study of the propeller of a boat

Today, the principle of the propeller has many applications in various fields, such as the activities of the Navy (which incidentally began using it before the aviation sector), wind power and nuclear power (the principle of the turbine is derived from the propeller). But it is used the most in the aviation sector, in various forms: classical propellers, turboprops and turbofans (composed of a series of compressors and turbines, derived from propellers). The water is driven around the propeller from the leading edge towards the trailing edge, creating a continuous screw effect. The water molecules are pressed against each other, causing them to push the boat forwards. Since shipbuilding first began, builders have always sought to improve performance and safety. Around a century ago, this desire prompted the nations with the most powerful naval presence to develop a basic set of tools for studying hydrodynamics and oceanic basins, which allowed a number of spectacular advancements in the areas of power prediction, maneuverability, hydroacoustics, vibroacoustics and seaworthiness.

	Ansys	Experimental
f_1	45.56	38
f_2	147.63	152
f_3	277.25	260
f_4	488.74	–
f_5	694.14	703

Table 3.4. *The first five natural frequencies of the immersed structure*

The propeller is the most important technical component of a ship. Its design and properties directly affect the energy yield. Using the wrong propeller will often lead to energy inefficiency. It is important to understand that designing a propeller is not straightforward, especially for fishing trawlers, whose technical specifications should be entrusted to a qualified and experienced expert. This kind of support can be given by propeller and engine manufacturers, and in some cases by the technical departments of governmental fishing development programs. What is the purpose of a propeller? This might seem like an obvious question – propellers transform the power supplied by the engine into thrust to propel the boat over the water. The design of the propeller needs to ensure that the boat will be effectively propelled forwards.

The four-blade propeller is the fastest type of propeller offered by Flexofold. It was designed and developed to meet the growing market demand for propellers to match the larger engines used by yachts, of around 55" to 65". With good reason, the Hanse 630 and Najad 570 are now equipped with four-blade propellers as standard. Looking at the profile and cross-section of the four-blade Flexofold, we see a powerful geometry that will effectively propel the boat, converting more power into thrust than any other propeller. And yet the propeller has very low drag, in part thanks to its thin and carefully designed hub; see Figure 4.2.

Figure 3.5. *Four-blade boat propeller*

3.5.1. *Numerical results*

To demonstrate the potential of the proposed method, we consider a concrete example from mechanics: a propeller submerged in water. Figures 3.6 and 3.7 present the finite element model of the structure.

The numerical calculations were performed with Ansys code. The proposed reduction method is applied to a simplified model of the propeller consisting of four substructures, each containing approximately the same number of elements (see Figure 3.7). The acoustic cavity is divided into four acoustic subdomains.

The material properties of both domains are as follows:

– for the structure: density = $9,200$ kg.m^{-3} ; Young's modulus = 9.6×10^{10} Pa; Poisson coefficient = 0.3.

– for the fluid: density = $1,000$ kg.m^{-3} ; speed of sound = $1,500$ m.s^{-1}.

Deterministic numerical calculations were performed for the structure as a whole, and then on a single blade both in air and water. The results are shown in Tables 3.5–3.7, and are compared with experimental results [MAN 13b]

Figure 3.8 shows the first three natural modes of a single blade in air and in water, and Figure 4.1 shows the first three modes of the full structure in air and water [MAN 13b].

Figure 3.6. *Finite element model of the full propeller. For a color version of this figure, see www.iste.co.uk/elhami/interactions.zip*

Figure 3.7. *Finite element model of the substructures. For a color version of this figure, see www.iste.co.uk/elhami/interactions.zip*

Modes	Ansys	Experimental
R_1	75.534	73
R_2	121.46	117
R_3	209.17	201

Table 3.5. *The first three natural frequencies of the full propeller in air*

Modes	Ansys	Experimental
R_1	37.581	36
R_2	68.629	65
R_3	127.46	123

Table 3.6. *The first three natural frequencies of the full propeller in water*

Modes	Blade in air	Blade in water
R_1	74.863	37.71
R_2	119.82	67.54
R_3	205.58	126.32

Table 3.7. *The first three natural frequencies of the blade, both in air and water, calculated with Ansys code*

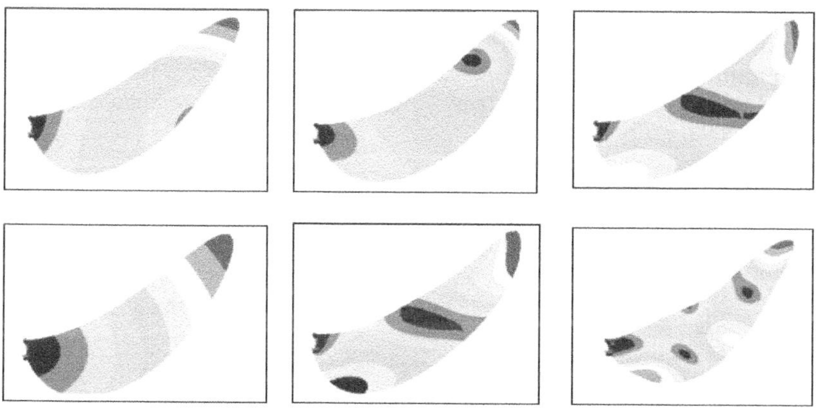

Figure 3.8. *The first three modes of the blade in air and water. For a color version of this figure, see www.iste.co.uk/elhami/interactions.zip*

The frequencies of the propeller are observed to be lower in water, which shows that the effects of the acoustic waves on the structure are correctly described by the numerical calculations. Tables 3.5 and 3.6 give the results of calculations and measurements in water and air performed for the full propeller, and Table 3.7 shows the results of calculations in water and air for a single propeller blade. The numerical calculations produce results comparable to those obtained in experiments, with uncertainty comparable to that quoted by other authors [SIG 11]. The model reduction step plays an important role in the numerical calculations for this problem. Indeed, we obtained a model with low degrees of freedom that is capable of reproducing the most important features of the full problem [RAD 15].

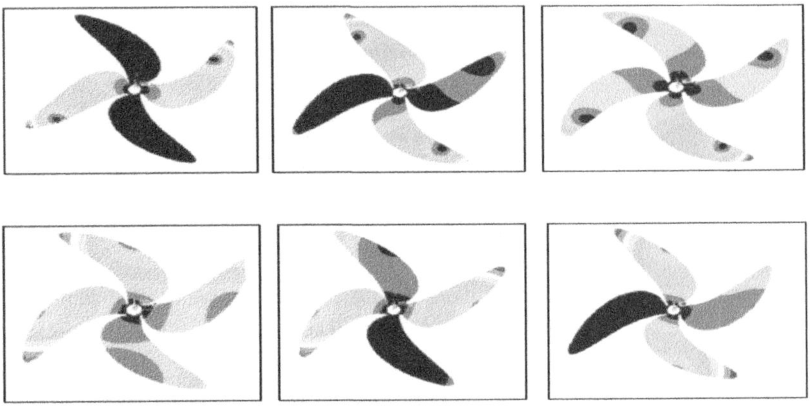

Figure 3.9. *The first three modes of the propeller in air and water. For a color version of this figure, see www.iste.co.uk/elhami/interactions.zip*

4

Aerodynamics

4.1. Introduction

This chapter presents the different numerical methods that can be used to solve fluid–structure interaction problems. Several of these methods date from the 1970s and have been regularly reused and refined ever since. The choice of method depends on the desired accuracy and on the strength of the coupling between the fluid and structure environments. In practice, using these numerical methods to solve coupled problems results in a nonlinear system of algebraic equations. This system can be solved with either a monolithic or a partitioned procedure.

We will present methods for discretizing the fluid and structure subproblems, followed by a list of coupling procedures. The final section illustrates the numerical application to the wing of an airplane that vibrates under the action of an air flow. Modal analysis is performed for the deformed wing subject to the resulting lift forces, allowing us to derive the natural modes and frequencies of the wing. The fluid flow is simulated using two different fluid dynamics solvers, Ansys/Flotran and Ansys/Fluent, enabling us to compare the results obtained in both of the cases of finite elements/finite elements and finite elements/finite volumes when processing the fluid/structure problem. We will also present another fluid–structure interaction problem within the context of transient analysis – a 3D plate oscillating within an air cavity subject to initial pressure applied to one side of the plate, which causes it to experience a deformation. Once this pressure is released, the plate oscillates back and forth as it recovers its equilibrium position. The fluid environment damps the motion of the plate, which reduces the amplitude of the oscillations over time.

4.2. Computational method

For many years now, modeling and numerically simulating the mechanics of problems with fluids coupled with immersed mobile obstacles has generated a great amount of scientific interest in a number of disciplines. Several different lines of research have been explored in order to improve the accuracy of solutions, facilitate their numerical implementations and reduce their computation times [HOU 12].

4.2.1. *Conformal mesh*

One idea for describing the interaction between an immersed, mobile geometry and the surrounding fluid is to use a conformal mesh. To do this, the edges of the obstacles are modeled precisely with a mesh. Triangular meshes are generally used to make it easier to generate elements near the edges, where they often need to be finer than the rest of the domain. These methods have the advantage that only the fluid domain needs to be discretized. The interaction with the obstacles unfolds at the interface via the application of boundary conditions. Figure 4.1 shows one example of this type of mesh.

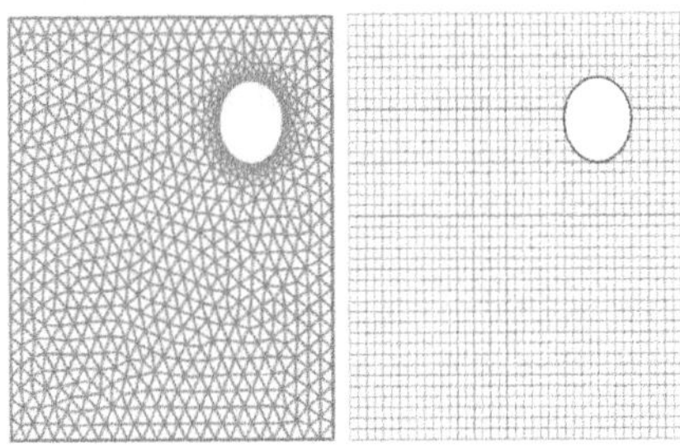

Figure 4.1. *Example of a conformal triangular and a non-conformal Cartesian mesh for a two-dimensional (2D) ellipse. For a color version of this figure, see www.iste.co.uk/elhami/interactions.zip*

The advantage of this method is that it is easy to implement when two solvers are available: one for the fluid equations and the other for the structure component. A coupling can then be constructed, and information can be exchanged through the interface.

If the obstacles are mobile, this method imposes a displacement on the mesh to ensure that the boundary of the fluid mesh coincides with the boundary of the obstacles. The mesh is deformed based on an arbitrary Lagrangian-Eulerian (ALE) formulation developed in the 1970s [HIR 74].

Many improvements have since been published, and Hu *et al.* [HU 01] gives a summary of the key results.

When the mesh is moved and the elements are deformed, the mesh needs to be regenerated at regular intervals. This procedure can prove expensive if the simulations are using a high resolution, especially in three dimensions. Furthermore, to achieve a good approximation of the boundary of the domain, we need to use small elements near the interface, especially in situations where we are interested in studying the contacts between obstacles [MAU 99]. This places constraints on the stability conditions of the numerical schemes that will later be used to perform integration. This method of generating the interface must also be combined with methods that are specifically suitable for the chosen type of mesh, namely the finite element method or the finite volume method. This means that we will need to put together a linear system, and the contributions of the small elements will degrade the conditioning of the linear system as a whole. Each time that the mesh is adjusted, these linear systems need to be reassembled. Solving these systems becomes difficult in two dimensions, and due to the size of the problems, other methods are required in three dimensions. This is the background in which fictitious domain methods have been studied.

4.2.2. *Immersed boundary methods*

The immersed boundary method was introduced by Peskin [PES 77] to model how blood flows in the heart interact with its elastic walls. Since this first study, several others have contributed by improving the method. Summaries of these results have been published by Peskin [PES 03] and Mittal and Iaccarino [MIT 05].

The method uses a set of Langrangian particles to describe the fictitious domain: the surface of the immersed obstacles, and then solves the fluid equations based on an Eulerian framework with a fixed, non-conformal mesh (Figure 4.2). On this surface, each particle carries a force describing the motion of the solid–fluid interface. This force is usually calculated from behavioral principles in structural mechanics (elasticity or rigid motion).

The forces carried by these particles are then used as the right-hand side in the momentum equation. Their contribution is represented as a sum of Dirac masses with the positions of each particle. Regularization techniques are used to numerically distribute these Dirac masses over the nodes of the Eulerian mesh, which are distinct from the particles.

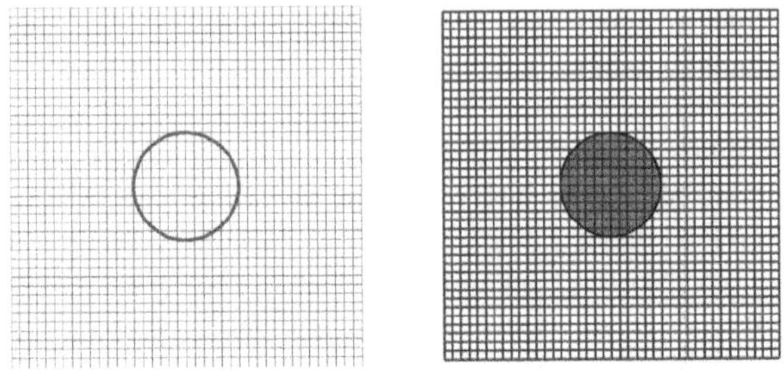

Figure 4.2. *Fictitious domain methods: difference between the immersed surface method on the left and the volume-based fictitious domain method on the right, in the case of a 2D disk. For a color version of this figure, see www.iste.co.uk/elhami/ interactions.zip*

After calculating the fluid velocity, the interface can be moved. The key step in this method is when the force terms of the Lagrangian particles are transferred onto the mesh. To do this, we must choose how to regularize the Dirac mass. Generally, this involves distributing the interface over multiple cells of the mesh [PON 09].

In [LEV 94], a generalization known as the immersed interface method is used to solve this problem by directly injecting the jumps in the normal derivatives into a finite difference scheme with adjusted coefficients near the interface. In [BOF 03], the Dirac masses are directly included in the variational formulation together with a finite element discretization.

4.2.3. *Volume-based fictitious domain methods*

The class of volume-based fictitious domain methods encompasses a large number of different methods for describing the interaction between the fluid and immersed obstacles. One aspect shared by all of these methods is that they do not use conformal meshes, but instead consider the interaction through a solid domain $B(t)$ that changes over time. These methods consider obstacles as volumes and not as surfaces like immersed boundary methods (Figure 4.2). This method was incidentally introduced as a way of extending the immersed boundary method to the rest of the volume of an obstacle. In [GLO 94], the authors introduce Lagrange multipliers to model a Dirichlet condition on the boundaries of the obstacles.

Remaining in the context of rigid solids [PAT 00], a rigidity constraint is applied within the obstacles using a Lagrange multiplier acting as a volume force. Finally, one other method was formulated based on a method for projecting onto rigid motions [PAT 01, SHA 05].

Another method was introduced in [ANG 99]: the penalization method. This method allows us to effectively handle Dirichlet conditions inside obstacles. To do this, a penalty term is added to the equation of conservation of momentum, but only to the nodes located in the fictitious volume domain of the solid. The advantage of this method is that it only requires the velocity field and characteristic functions of the obstacles to be known, which makes it easy to implement.

4.3. Aerodynamic problem's resolution

4.3.1. *Mobile domain*

In the case where the domain Ω_f can undergo deformation or motion over time, we will choose the ALE formulation [BOF 03, DON 82, FOR 04]. Usually, when the fluid domain is fixed, the spatial or Eulerian formulation is chosen, which is sufficient to describe the flow. In this case, we are interested in one or more of the quantities associated with the fluid (such as the speed or the pressure) at a given position or within a given domain (such as Ω_f). The Lagrangian or material formulation is more commonly used when the domain can deform and move, while continuing to be phrased in terms of the initial configuration. Since the Lagrangian formulation is more typical for the solid, one formulation capable of uniting the two coexisting aspects of fluid-solid interaction is given by ALE. This formulation combines the regions in the fluid domain where we need to approximate a Lagrangian formulation to follow the motion of particles (for example near an interface between the fluid and the solid) and the regions where an Eulerian description is sufficient to describe the flow. We therefore require a reference configuration Ω_0^f, for which we use the Lagrangian coordinates X. The domain Ω_t^f is called the updated configuration with Eulerian coordinates x. Thus, we can define the map A_t sending the reference configuration to the updated configuration as

$$\left\{\begin{array}{rl} \mathcal{A}_t : \Omega_0^f \subset \mathbb{R} & \rightarrow \ \Omega_t^f \subset \mathbb{R}^d \\ \mathbf{X} & \rightarrow \ x(\mathbf{X}, \mathbf{t}) = \mathcal{A}_\mathbf{t}(\mathbf{X}), \forall \mathbf{t} \in \mathbf{I} \end{array}\right. \quad [4.1]$$

The integer d is the dimension of the space and $I = [t_0, T]$ is the time interval of integration. Thus, we define the ALE velocity w.

$$w = \left.\frac{\partial \mathcal{A}_t}{\partial t}\right|_X = \left.\frac{\partial x(\mathbf{X}, t)}{\partial t}\right|_X \qquad [4.2]$$

The derivative of v with respect to (w.r.t.) time in this configuration may then be written as:

$$\left.\frac{\partial v}{\partial t}\right|_X = \frac{\partial v}{\partial t} + w \cdot \nabla v \qquad [4.3]$$

We will later use the w field to calculate a velocity for the mesh. Interpreting equation [4.3] allow us to identify a number of different situations. If the speed w is equal to v, we recover the definition of the material derivative of the velocity, and if it is equal to zero, we obtain the Eulerian derivative of the velocity [DON 82, DON 03]. This explains why the configuration constructed by this approach is given the name of ALE. It allows us to locally switch between the Lagrangian and Eulerian configurations within the same geometry. When the gradient operator ∇ is expressed in terms of the coordinates X, we write ∇X. Incorporating [4.3] into [3.82] gives the ALE formulation of the Navier–Stokes equation:

$$\left.\frac{\partial v}{\partial t}\right|_X + (v - w) \cdot \nabla v = -\nabla p + \frac{1}{Re} \nabla \cdot d(v) + b_f \qquad [4.4]$$

In this formulation, the system to be solved is given by equations [4.4] and [3.83].

4.3.2. *Weak formulation*

The weak formulation reduces the order of the partial derivatives using integration by parts. This allows us to expand the spaces in which the numerical solution is constructed. We will therefore define various Sobolev spaces needed to express the weak formulations, which are for the most part subsets of the Hilbert space H^1 and the Lebesgue space L^2 [QUA 94] for an arbitrary domain $\Omega \in \mathbb{R}$. These spaces are necessary to meet the hypotheses of the Lax–Milgram theorem [DEV 02], which guarantees the existence and uniqueness of a solution in the weak formulations that we will consider. The space $L^2(\Omega)$ is defined as the following set:

$$L^2(\Omega) = \left\{ u; \text{ measurable scalar function} : \int_\Omega |u(x)|^2 \, d\Omega < \infty \right\} \quad [4.5]$$

For a given function $u \in L^2(\Omega^f)$, this space is equipped with the norm

$$\|u\|_{L^2(\Omega)} = \left(\int_\Omega (|u(x)|^2 \, d\Omega \right)^{\frac{1}{2}} \quad [4.6]$$

The Sobolev space $H^1(\Omega)$ is the following set of functions:

$$H^1(\Omega) = \left\{ u \in L^2(\Omega) : \frac{\partial u}{\partial x_k} \in L^2(\Omega), \forall k \in [1, d] \cap \mathbb{N} \right\} \quad [4.7]$$

This space is equipped with the norm:

$$\|u\|_{H^1(\Omega)} = \left(\int_\Omega \left(|u(x)|^2 + \sum_{k=1}^{d} \left| \frac{\partial u}{\partial x_k} \right|^2 \right) d\Omega \right)^{\frac{1}{2}} \quad [4.8]$$

In the fluid

To write the weak formulation of [4.4], we first introduce the space $H^1_{0,D}(\Omega^f_0)^d$:

$$H^1_{0,D}(\Omega^f_0)^d = \left\{ u \in H^1_{0,D}(\Omega^f_0)^d : u|_{\Gamma_D} = 0 \right\} \quad [4.9]$$

The test functions $\hat{\mathbf{v}}$ that will allow us to establish the weak formulation of [4.4] are then chosen from $H^1_{0,D}(\Omega^f_0)^d$. Henceforth, we have: $\hat{\mathbf{v}} \in H^1_{0,D}(\Omega^f_0)^d$.

A relation can be found between the test functions on Ω^f_0 and the same test functions viewed on Ω^f_t, denoted $\mathbf{v}$, by using [4.1]

$$\mathbf{v}(x) = \mathbf{v}(\mathcal{A}_t(X)) = \hat{\mathbf{v}}(X) = \hat{\mathbf{v}} \circ \mathcal{A}_t^{-1}(x) \quad [4.10]$$

This allows us to write an expression for the weak formulation, while emphasizing that the test functions change as a function of the motion of the domain over time. We now define the space of test functions $\mathbf{v}$ from the space $\hat{\mathbf{v}}$ as:

$$V(\Omega_t^f)^d = \left\{\mathbf{v} : \Omega_t^f \times I \to \mathbb{R}^d, \mathbf{v}(x) = \hat{\mathbf{v}} \circ \mathcal{A}_t^{-1}(x), \hat{\mathbf{v}} \in H^1_{0,D}(\Omega_0^f)^d\right\} \quad [4.11]$$

Similarly, we seek to find the velocity v in the following space:

$$V_D(\Omega_t^f)^d = \left\{u \in V(\Omega_t^f)^d : u|_{\Gamma_D} = \bar{v}|_{\Gamma_D}\right\} \quad [4.12]$$

The weak formulation in the ALE configuration can be written using [4.4] and [3.83]:

$$\begin{cases} \text{Find } (v,p) \in V_D(\Omega_t^f)^d \times L^2(\Omega_t^f),\ t \in I \text{ such that} \\ \displaystyle\int_{\Omega_t^f} (\frac{\partial v}{\partial t}|_x + (v-w)\cdot\nabla v)\cdot \mathbf{v} d\Omega = \int_{\Omega_t^f} [p\nabla\cdot\mathbf{v} - \frac{2}{Re}\nabla v : d(\mathbf{v}) \\ \qquad\qquad + b_f \cdot \mathbf{v}] d\Omega \quad \forall \mathbf{v} \in V(\Omega_t^f)^d \\ \displaystyle\int_{\Omega_t^f} (\nabla\cdot v)\mathbf{q} d\Omega = 0 \quad \forall \mathbf{q} \in L^2(\Omega_t^f) \end{cases} \quad [4.13]$$

In the solid

Unlike the fluid part, the solid domain is calculated from a Lagrangian perspective. The displacement field u added to the initial position Ω_0^s of the solid domain gives the current position of the solid Ω_t^s. Thus, we seek to find u in the following space:

$$W(\Omega_t^s)^d = \{v \in H^1(\Omega_t^s)^d : v|_{\Gamma_D^s} = \bar{u}|_{\Gamma_D^s}\} \quad [4.14]$$

The test function $\mathbf{v}_s$ associated with the structure solution is therefore chosen from

$$W_0(\Omega_t^s)^d = \{v \in H^1(\Omega_t^s)^d : v|_{\Gamma_D^s} = 0\} \quad [4.15]$$

The weak formulation that we will solve inside the structure may therefore be written as:

$$\begin{cases} \text{Find } u \in W(\Omega_t^s)^d \\ \displaystyle\int_{\Omega_t^s} \sigma_s : \nabla \mathbf{v}_s d\Omega + \int_{\Omega_t^s} \rho_s \frac{\partial^2 u}{\partial t^2} \cdot \mathbf{v}_s d\Omega = \int_{\Gamma_N^s(t)} \sigma_s n_s \cdot \mathbf{v}_s d\Gamma + \\ \displaystyle\qquad \int_{\Omega_t^s} \rho_s b_s \cdot \mathbf{v}_s d\Omega \quad \forall \mathbf{v}_s \in W_0(\Omega_t^s)^d \end{cases} \quad [4.16]$$

At the interface

The boundary conditions that we will apply to the fluid–structure interface reference the surface stress forces arising from the fluid. They represent a source term on the surface of the solid that is highly significant in computing the dynamics. The weak formulation on the interface $\Gamma_N^s(t)$ uses the test functions ϑ, which are simply the restrictions of the test functions $\mathbf{v}_s$ to $\Gamma_N^s(t)$:

$$\vartheta = \mathbf{v}_s|_{\Gamma_N^s(t)} \quad [4.17]$$

Interpolating [3.91] with the weak formulation gives

$$\begin{aligned} \int_{\Gamma_N^s(t)} \sigma_s n_s \cdot \vartheta d\Gamma &= -\int_{\Gamma_N^s(t)} \sigma_f n \cdot \vartheta d\Gamma \\ &= \int_{\Gamma_N^s(t)} [pIn - \mu_f(\nabla v + \nabla v^T)n] \cdot \vartheta d\Gamma \end{aligned} \quad [4.18]$$

4.3.3. *Evaluating the energy of the system*

We use the Eulerian perspective in the fluid to determine the energy of the Navier Stokes equations. The system energy can thus be evaluated by taking the scalar product of the Navier–Stokes equations [3.80] by v. We then integrate these equations over space, and once again over time [GRA 12]. Integrating by parts on Ω_t^f, we thus obtain

$$\int_{\Omega_t^f} \rho_f(\frac{\partial v}{\partial t} + v \cdot \nabla v + \nabla p - \nu_f \Delta v - b_f) \cdot v d\Omega = 0 \Rightarrow$$

$$\rho_f \int_{\Omega_t^f} (\frac{\partial v}{\partial t} \cdot v + (v \cdot \nabla v) \cdot v - p\nabla \cdot v + \nu_f \nabla v : \nabla v - b_f \cdot v) d\Omega$$

$$= \int_{\Gamma_t^f} \sigma_f n \cdot v d\Gamma$$

We then expand the integral on the boundary Γ_t^f defined as follows:

$$\Gamma_t^f = \Gamma_{\text{wall}} \cup \Gamma_{\text{in}} \cup \Gamma_I \cup \Gamma_{\text{out}} \quad [4.19]$$

Each of these boundaries are illustratively shown in Figure 4.3, which shows a solid immersed in a fluid domain.

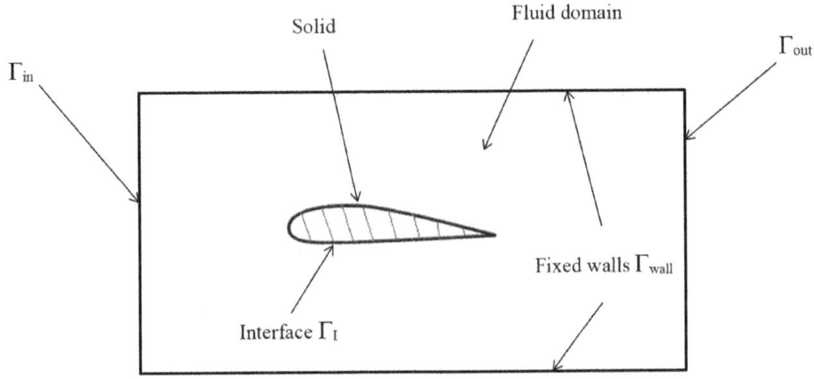

Figure 4.3. *Geometry of a solid immersed in a fluid*

On Γ_{in}, some types of flow will have conditions imposed on the velocity, whereas on Γ_{wall} this velocity will be assumed to be zero. At the interface Γ_I, the boundary conditions are defined according to [3.90] and [3.91], whereas a stress-free condition is imposed on Γ_{out} denoting a fluid outlet. Thus, we have

$$\begin{aligned}\int_{\Gamma_t^f} \sigma_f n \cdot v d\Gamma &= \int_{\Gamma_I} \sigma_f n \cdot v d\Gamma + \int_{\Gamma_{\text{wall}}} \sigma_f n \cdot v d\Gamma + \int_{\Gamma_{\text{in}}} \sigma_f n \cdot v d\Gamma \\ &\quad + \int_{\Gamma_{\text{out}}} \sigma_f n \cdot v d\Gamma \\ &= \int_{\Gamma_{\text{I}}} \sigma_f n \cdot v d\Gamma + \int_{\Gamma_{\text{in}}} \sigma_f n \cdot v d\Gamma \end{aligned} \quad [4.20]$$

From the equality $(v \cdot \nabla v) \cdot v = \frac{1}{2}\nabla(v \cdot v) \cdot v$ and incompressibility, we can simplify the convective term

$$\int_{\Omega_t^f} (v \cdot \nabla v) \cdot v d\Omega = \frac{1}{2} \int_{\Omega_t^f} \nabla(v \cdot v) \cdot v d\Omega$$

$$= \frac{1}{2} \left[- \int_{\Omega_t^f} (v \cdot v)(\nabla \cdot v) d\Omega + \int_{\Gamma_t^f} (v \cdot v)(v \cdot n) d\Gamma \right] \Rightarrow$$

$$\int_{\Omega_t^f} (v \cdot \nabla v) \cdot v d\Omega = \frac{1}{2} \int_{\Gamma_t^f} (v \cdot v)(v \cdot n) d\Gamma$$

By the Reynolds transport theorem, we can express the Eulerian derivative of the velocity as follows:

$$\int_{\Omega_t^f} \frac{\partial v}{\partial t} \cdot v d\Omega = \frac{1}{2} \int_{\Omega_t^f} \frac{\partial (v \cdot v)}{\partial t} d\Omega$$

$$= \frac{1}{2} \left[\frac{d}{dt} \int_{\Omega_t^f} (v \cdot v) d\Omega - \int_{\Gamma_t^f} (v \cdot v)(v \cdot n) d\Gamma \right] \quad [4.21]$$

The integral term with the pressure is equal to zero by incompressibility. In summary, in the fluid domain we have

$$\rho_f \frac{1}{2} \frac{d}{dt} \int_{\Omega_t^f} v \cdot v d\Omega + \rho_f \int_{\Omega_t^f} \nu_f \nabla v : \nabla v - b_f \cdot v d\Omega$$

$$= \int_{\Gamma_I} \sigma_f n \cdot v d\Gamma + \int_{\Gamma_{\text{in}}} \sigma_f n \cdot v d\Gamma \quad [4.22]$$

In the structure, we integrate equation [3.85] on Ω_t^s and multiply by $\dot{u}$, then perform integration by parts to obtain

$$\int_{\Omega_t^f} \rho_s \frac{\partial^2 u}{\partial t^2} \cdot \dot{u} - (\nabla \cdot \sigma_s) \cdot \dot{u} - \rho_s b_s \cdot \dot{u} d\Omega = 0 \Rightarrow$$

$$\frac{1}{2} \frac{d}{dt} \int_{\Omega_t^s} \rho_s \dot{u} \cdot \dot{u} d\Omega + \int_{\Omega_t^s} (\lambda(\nabla \cdot u)(\nabla \cdot \dot{u}) + 2\mu\varepsilon(u) : \varepsilon(\dot{u}) - \rho_s b_s \dot{u}) d\Omega$$

$$= \int_{\Gamma_I} \sigma_s n_s \cdot \dot{u} d\Gamma \quad [4.23]$$

As with u in equation [1.38], we also have

$$\varepsilon(\dot{u}) = \frac{1}{2}(\nabla \dot{u} + (\nabla \dot{u})^T) \qquad [4.24]$$

Furthermore, we have

$$\int_{\Gamma_I} \sigma_s n_s \cdot \dot{u} d\Gamma = -\int_{\Gamma_I} \sigma_s n \cdot \dot{u} d\Gamma \qquad [4.25]$$

We replace [4.25] in [4.23]. We then make the following approximation to isolate the derivative w.r.t. time of the integral in [4.25]:

$$\frac{1}{2}\frac{d}{dt}\left[\int_{\Omega_t^s} \rho_s \dot{u} \cdot \dot{u} + \lambda(\nabla \cdot u)^2 + 2\mu\varepsilon(u) : \varepsilon(u) d\Omega\right] - \int_{\Omega_t^s} \rho_s b_s \cdot \dot{u} d\Omega = -\int_{\Gamma_I} \sigma_s n_s \cdot \dot{u} d\Gamma \qquad [4.26]$$

Equation [4.26] is a simplification that will make it easier to evaluate these integrals with respect to time. From [1.45] and [1.48], we also have

$$\int_{\Gamma_I} \sigma_s n \cdot \dot{u} d\Gamma = \int_{\Gamma_I} \sigma_f n \cdot v d\Gamma \qquad [4.27]$$

Substituting [4.27] into [4.22], we obtain

$$\rho_f \frac{1}{2}\frac{d}{dt}\int_{\Omega_t^f} v \cdot v d\Omega + \rho_s \frac{1}{2}\frac{d}{dt}\int_{\Omega_t^f} \dot{u} \cdot \dot{u} d\Omega + \rho_f \int_{\Omega_t^f} \nu_f \nabla v : \nabla v d\Omega +$$
$$\frac{1}{2}\frac{d}{dt}\left[\int_{\Omega_t^s} \lambda(\nabla \cdot u)^2 + 2\mu\varepsilon(u) : \varepsilon(u) d\Omega\right]$$
$$= \rho_f \int_{\Omega_t^f} b_f \cdot v d\Omega + \rho_s \int_{\Omega_t^s} b_s \cdot \dot{u} d\Omega + \int_{\Gamma_{in}} \sigma_f n \cdot v d\Gamma \qquad [4.28]$$

We can now integrate [4.28] with respect to time to deduce the energy of the system

$$\int_0^T \frac{dE(t)}{dt} dt = \rho_f \frac{1}{2}\int_0^T \frac{d}{dt}\int_{\Omega_t^s} v \cdot v d\Omega dt + \rho_s \frac{1}{2}\int_0^T \frac{d}{dt}\int_{\Omega_t^s} \dot{u} \cdot \dot{u} d\Omega dt$$

$$+\rho_f \int_0^T \int_{\Omega_t^f} \nu_f \nabla v : \nabla v d\Omega dt + \frac{1}{2}\int_0^T \frac{d}{dt}\int_{\Omega_t^s} \lambda(\nabla \cdot u)^2$$

$$+2\mu\varepsilon(u) : \varepsilon(u) d\Omega dt - \rho_f \int_0^T \int_{\Omega_t^s} b_f \cdot v d\Omega dt - \rho_s \int_0^T \int_{\Omega_t^s} b_s \cdot \dot{u} d\Omega dt$$

$$-\int_0^T \int_{\Gamma_{in}} \sigma_f n \cdot v d\Gamma dt = 0 \Rightarrow E(T) - E(0) \quad [4.29]$$

$$= \rho_f \frac{1}{2}\int_{\Omega_t^s} (v \cdot v)(T) d\Omega + \rho_s \frac{1}{2}\frac{d}{dt}\int_{\Omega_t^s} (\dot{u} \cdot \dot{u})(T) d\Omega$$

$$+\rho_f \int_0^T \int_{\Omega_t^f} \nu_f \nabla v : \nabla v d\Omega dt + \frac{1}{2}\int_{\Omega_t^s} \lambda(\nabla \cdot u(T))^2$$

$$+2\mu\varepsilon(u(T)) : \varepsilon(u(T)) d\Omega - \rho_f \int_0^T \int_{\Omega_t^s} b_f \cdot v d\Omega dt$$

$$-\rho_s \int_0^T \int_{\Omega_t^s} b_s \cdot \dot{u} d\Omega dt - \int_0^T \int_{\Gamma_{in}} \sigma_f n \cdot v d\Gamma dt$$

$$-\rho_f \frac{1}{2}\int_{\Omega_0^s} (v \cdot v)(0) d\Omega - \rho_s \frac{1}{2}\int_{\Omega_0^s} (\dot{u} \cdot \dot{u})(0) d\Omega - \frac{1}{2}\int_{\Omega_0^s} \lambda(\nabla \cdot u(0))^2$$

$$+2\mu\varepsilon(u(0)) : \varepsilon(u(0)) d\Omega = 0$$

This final equation expresses the conservation of energy.

Assuming that the initial conditions are $u_0 = \dot{u}_0 = v_0$ and neglecting b_f and b_s, we simplify [4.30] by

$$\rho_f \frac{1}{2}\int_{\Omega_t^s} (v \cdot v)(T) d\Omega + \rho_s \frac{1}{2}\frac{d}{dt}\int_{\Omega_t^s} (\dot{u} \cdot \dot{u})(T) d\Omega + \rho_f \int_0^T \int_{\Omega_t^f} \nu_f \nabla v :$$

$$\nabla v d\Omega dt + \frac{1}{2}\int_{\Omega_t^s} \lambda(\nabla \cdot u(T))^2 + 2\mu\varepsilon(u(T)) :$$

$$\varepsilon(u(T)) d\Omega - \int_0^T \int_{\Gamma_{in}} \sigma_f n \cdot v d\Gamma dt \quad [4.30]$$

The first two terms of equation [4.30] represent the kinetic energy of the fluid and the solid. The third term corresponds to the energy dissipated in the fluid by viscoscity, and the fourth term is the elastic potential energy of the solid. All of this energy is therefore primarily determined by the flow imposed on the boundary Γ_I.

4.3.4. *Numerically solving the system*

The equations that must be solved were presented above. They are expressed using an ALE formulation. We will now give a description of the tools available for solving the system. We will discuss the choice of temporal discretization. We will then outline the method for the solving the system with respect to time, and the tools available for processing the nonlinear term in the equation.

Choice of temporal discretization

To reduce the number of time steps, we choose a second-order implicit scheme. We will consider the backward Euler method of order 2 (BE2, also known as the backward differential formula of order 2). This is equivalent to discretizing the temporal acceleration of the momentum equation as follows:

$$\frac{\partial v}{\partial t} = \frac{3v^{n+1} - 4v^n + v^{n-1}}{2\Delta t} \qquad [4.31]$$

while handling the other terms of the equation implicitly, where $n+1$, n and $n-1$, respectively, denote the values of quantities at each successive time step, and $n+1$ is the current time step.

Solving the system with respect to time

Here, we will outline the method used to solve the system with respect to time.

$$\left\{ \begin{array}{c} \nabla \cdot v^{n+1} = 0 \\ \left(\frac{3v^{n+1} - 4v^n + v^{n-1}}{2\Delta t}\right) + (v^{n+1} - w) \cdot \nabla v^{n+1} = -\nabla p^{n+1} + \frac{1}{Re}\Delta v^{n+1} \end{array} \right. \qquad [4.32]$$

The chosen BE2 scheme is unconditionally stable, but solving it requires inverting the system, which is computationally highly expensive. To reduce the time required, we use a projection method known as the Incremental Pressure-Correction Scheme in Standard Form [GUE 06]. This method has three steps:

– An initial approximation step in which the momentum equation of the system [4.32] is solved using the value of the pressure field at the previous time step. This gives a velocity field v^* that satisfies

$$\left(\frac{3v^{n+1} - 4v^n + v^{n-1}}{2\Delta t}\right) + (v^* - w) \cdot \nabla v^* = -\nabla p^{n+1} + \frac{1}{Re}\Delta v^* \qquad [4.33]$$

The equation of conservation of momentum of system [4.32] is subtracted from equation [4.33]. The differences between the convective acceleration terms and the Laplacian velocity terms are of the same order as the truncation error [VUI 11]. Hence,

$$\frac{3}{2\Delta t}(v^{n+1} - v^*) + \nabla(p^{n+1} - p^n) = 0 \qquad [4.34]$$

– The second step projects [4.34] onto the velocity space with zero divergence, which is equivalent to applying the divergence operator to [4.34] in order to calculate the updated pressure field. After projecting, this equation becomes

$$\Delta p^{n+1} = \frac{3}{2\Delta t}\nabla \cdot v^* - \Delta p^n \qquad [4.35]$$

– Once the updated pressure field is known, the velocity field v^{n+1} may be deduced from [4.34].

This method requires a temporal discretization scheme that is unconditionally stable and of the second order, which justifies our choice of the second-order backward Euler scheme. Moreover, it was shown in [GUE 06] that the precision of this method is at best of order 2, so there is no benefit in choosing a higher order discretization scheme.

Handling the nonlinear convection term

The Navier–Stokes equations are nonlinear with respect to the unknown considered here. This nonlinearity is located in the convection term. In the projection method, we must deal with this nonlinearity in order to determine the velocity field v^* in equation [4.33]. We will describe the methods available to manage this nonlinearity.

Newton's method

One way of dealing with the nonlinearity is Newton's method. To illustrate this method, consider the following one-dimensional (1D) steady convection–diffusion problem:

$$v \cdot \nabla v = \frac{1}{Re} \Delta v \qquad [4.36]$$

Newton's method is an iterative method. The goal is to find the zero of the residue $R(v)$ by successively subiterating.

$$R(v) = v \cdot \nabla v - \frac{1}{Re} \Delta v = 0 \qquad [4.37]$$

We linearize this residue to give $R_{lin}(v)$ around an initial state v_i, where i is the subiteration index $(0 \leq i \leq I_{\max})$, such that:

$$R_{lin}(v) = \delta_{v-v_i} R(v_i) + R(v_i) = 0 \qquad [4.38]$$

where $\delta_g R(f)$ is the derivative in the sense of operators of R at f with respect to g and is given by:

$$\delta_g R(f) = \lim_{\epsilon \to 0} \frac{d}{d\epsilon} R(f + \epsilon g) \qquad [4.39]$$

In the convection–diffusion equation, we obtain:

$$\delta_g R(f) = ((g \cdot \nabla) f + (f \cdot \nabla) g) - \frac{1}{Re} \Delta g \qquad [4.40]$$

Since $\delta_g R(f)$ was linearized with respect to g, equation [4.38] is equivalent to finding:

$$\delta_v R(v_i) + R(v_i) - \delta_{v_i} R(v_i) = 0 \qquad [4.41]$$

or in other words:

$$v \cdot \nabla v_i + v_i \cdot \nabla v - \frac{1}{Re} \Delta v = v_i \cdot \nabla v_i \qquad [4.42]$$

Solving this system yields an estimate v_{i+1} of the solution. If this estimate is not satisfactory, the residue is once again linearized at v_{i+1}. In practice, we want $\|v_{i+1} - v_i\| < \xi$ where ξ is the chosen convergence criterion. This method has the advantage of rapidly converging toward the desired solution when the initial estimate v_0 is sufficiently close. If the temporal discretization is relatively coarse, the solutions can vary strongly from one time step to the next. Fixed point methods with higher radii of convergence are therefore preferable.

Fixed point method

The principle of the fixed point method, also known as Picard's method, is to replace the derivative of the residue with an approximate estimate. Thus, in equation [4.40], we can approximate $\delta_g R(f)$ in two ways:

$$\delta_g R(f) = f \cdot \nabla g - \frac{1}{Re}\Delta g \qquad [4.43]$$

$$\text{or} \quad \delta_g R(f) = g \cdot \nabla f - \frac{1}{Re}\Delta g$$

This is equivalent to choosing to retain one of the two terms in the linearization of the convective term. The choice of which term to keep is made empirically. In our case, we will use the first expression of [4.43]. This is the one that produces better convergence in practice for this method. Thus, [4.38] becomes

$$v_i \cdot \nabla v - \frac{1}{Re}\Delta v = v_i \nabla v_i \qquad [4.44]$$

Similarly to Newton's method, we obtain an estimate v_{i+1} of the solution.

Although the fixed point method is less restrictive in the choice of initial estimate, it is still desirable to further decrease the dependence on this estimate. To do this, we apply a relaxation to the fixed point method.

Relaxation of the fixed point method

We introduce the relaxation parameter $0 < \lambda \leq 1$ and solve the system as for a fixed point. However, we instead obtain an intermediate estimate v_{int} where

$$v_{int} = \lambda v_{int} + (1 - \lambda) v_i \qquad [4.45]$$

Note that choosing $\lambda = 1$ recovers the fixed point method.

We have now presented the tools for solving the system with respect to time, and we chose to use the temporal BE2 scheme. The solution with respect to time is found using a projection–correction method whose nonlinearities are managed using a relaxed fixed point method. We will now discuss the spatial discretization of the system.

4.3.5. *Discretization by finite elements*

We must solve the following problem:

$$\begin{cases} \text{Find } (v,p) \in V_D(\Omega_t^f)^d \times L^2(\Omega_t^f),\ t \in I \text{ such that} \\ \displaystyle\int_{\Omega_t^f} (\frac{\partial v}{\partial t}|_x + (v-w)\cdot\nabla v)\cdot \mathbf{v}d\Omega \\ \quad = \displaystyle\int_{\Omega_t^f} [p\nabla\cdot\mathbf{v} - \frac{2}{Re}\nabla v : d(\mathbf{v}) + b_f\cdot\mathbf{v}]d\Omega \quad \forall \mathbf{v}\in V(\Omega_t^f)^d \\ \displaystyle\int_{\Omega_t^f} (\nabla\cdot v)\mathbf{q}d\Omega = 0 \quad \forall\mathbf{q}\in L^2(\Omega_t^f) \end{cases} \qquad [4.46]$$

Matrix expression

To discretize the problem, we apply the Galerkin method. We introduce the approximation spaces M_h and P_h, which are the subspaces of dimension $2K$ and L of the function spaces $V^1(\Omega^f)$ and $L^2(\Omega^f)$. Each of these two subspaces is defined by constructing bases of finite elements $\mathbf{v}_k(x)_{k=1}^{2K}$ and $\mathbf{q}_k(x)_{k=1}^{L}$. The dimensions K and L depend on the chosen type of finite elements and the number of elements used to create the mesh. The first step is to establish the matrix system resulting from system [4.46], after which we will list the chosen types of finite element.

We approximate the variables v and p by:

$$\begin{cases} v_h(x) = \sum\limits_{k=1}^{2K} v_k\mathbf{v}(x) \\ p_h(x) = \sum\limits_{k=1}^{L} p_k\mathbf{q}(x) \end{cases} \qquad [4.47]$$

and we replace them by their approximations in the system ([4.13] and [4.46]). The problem to be solved holds for any pair of test functions $(\mathbf{v}, \mathbf{q})$. It therefore holds for any pair of basis functions $(\mathbf{v_k}, \mathbf{q}_l)_{k=1\ldots 2K, l=1\ldots L}$. If one such pair is given, the problem ([4.13] and [4.46]) can be approximated by:

$$\begin{cases} \text{Find } (v_h, p_h) \in M_H \times P_H \text{ such that} \\ \displaystyle\sum_{i=1}^{2K} v_i \int_{\Omega^f} \mathbf{q}_l \nabla \cdot \mathbf{v_i} d\Omega = 0 \\ \displaystyle\sum_{i=1}^{2K} \frac{\partial v_i}{\partial t} \int_{\Omega^f} \mathbf{v_k} \cdot \mathbf{v}_i d\Omega + \sum_{i=1}^{2K} v_i \int_{\Omega^f} \mathbf{v}_k \cdot (v - w) \cdot \nabla \mathbf{v}_i d\Omega = \\ \displaystyle\sum_{j=1}^{L} p_j \int_{\Omega^f} \mathbf{q}_l \nabla \cdot \mathbf{v}_k d\Omega - \frac{1}{Re} \sum_{i=1}^{2K} v_i \int_{\Omega^f} \nabla \mathbf{v}_k : (\nabla \mathbf{v}_i + {}^t\nabla \mathbf{v}_i) d\Omega \\ \forall (\mathbf{v}_k, \mathbf{q}_l)_{k=1\ldots 2K, l=1\ldots L} \in M_H \times P_H \end{cases} \quad [4.48]$$

The system of equation [4.48] written in matrix form gives:

$$\begin{cases} \mathcal{G} V = 0 \\ \mathcal{M} \dfrac{\partial V}{\partial t} + \mathcal{N}(v - w) V + {}^t\mathcal{G} P + \dfrac{1}{Re} \mathcal{L} V = 0 \end{cases} \quad [4.49]$$

where

$$\begin{cases} V = v_i = {}^t(v_1, \ldots, v_{2K}) \\ P = p_j = {}^t(p_1, \ldots, p_L) \\ \mathcal{M} = m_{i,j} = \displaystyle\int_{\Omega^f} \mathbf{v}_j \cdot \mathbf{v}_i d\Omega \\ \mathcal{N}(\bullet) = n_{i,j}(\bullet) = \displaystyle\int_{\Omega^f} \mathbf{v}_j \cdot ((\bullet) \cdot \nabla) \mathbf{v}_i d\Omega \\ \mathcal{G} = g_{i,j} = -\displaystyle\int_{\Omega^f} \mathbf{q}_j \nabla \cdot \mathbf{v}_i d\Omega \\ \mathcal{L} = l_{i,j} = \displaystyle\int_{\Omega^f} \nabla \mathbf{v}_j : (\nabla \mathbf{v}_i + {}^t\nabla \mathbf{v}_i) d\Omega \end{cases} \quad [4.50]$$

To solve this mixed problem with finite elements, the approximation spaces M_h and P_h must satisfy a compatibility condition often called the *Inf-Sup* or Ladysenskaya–Babuska–Brezzi) condition [FAR 98]. Because of this, we are not free to choose any type of finite element.

There are two major classes of finite elements in fluid dynamics computations: elements that allow a continuous approximation of the pressure and elements that allow a discontinuous approximation of the pressure. In the first class, the most popular elements are the Taylor–Hood element $\mathbb{P}_2/\mathbb{P}_1$ and its variants $\mathbb{Q}_2/\mathbb{Q}_1$, $\mathbb{P}_k/\mathbb{P}_{k1}$ and $\mathbb{Q}_k/\mathbb{Q}_{k1}$, $k \geq 2$ (Figure 4.4). This element is quadratic in the velocity and linear in the pressure. Its order of convergence in space is $O(h^3)$ for the velocity and $O(h^2)$ for the pressure.

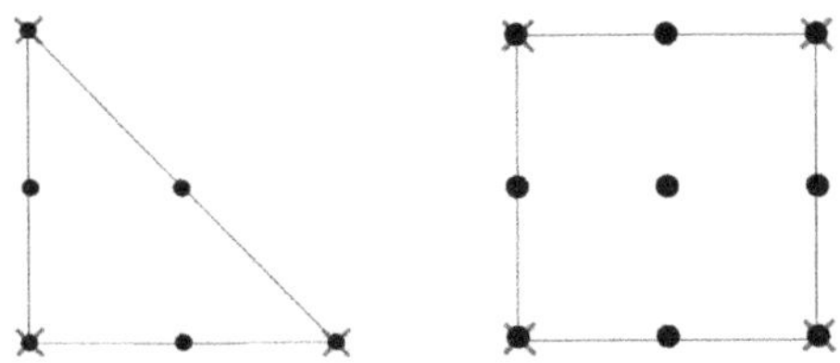

(a) $\mathbb{P}_2/\mathbb{P}_1$ (b) $\mathbb{Q}_2/\mathbb{Q}_1$

Figure 4.4. *Taylor–Hood elements: • velocity field nodes, × pressure field nodes. For a color version of this figure, see www.iste.co.uk/elhami/interactions.zip*

In the second class, the most commonly used elements are the Crouzeix–Raviart element $\mathbb{P}_2$-bubble/$\mathbb{P}_1$-discontinuous, its variants $\mathbb{Q}_2/\mathbb{P}_1$-discontinuous and its generalization to hexahedra. This element is equivalent to the Taylor–Hood element in terms of order of convergence [FAR 98]. It is sometimes even better, since the velocity estimate is enhanced by a bubble function associated with the center of the element (Figure 4.5).

The algorithms used for fluid–structure interaction require the methods used for the fluid and the structure to be stable, and also require the coupling scheme to be stable.

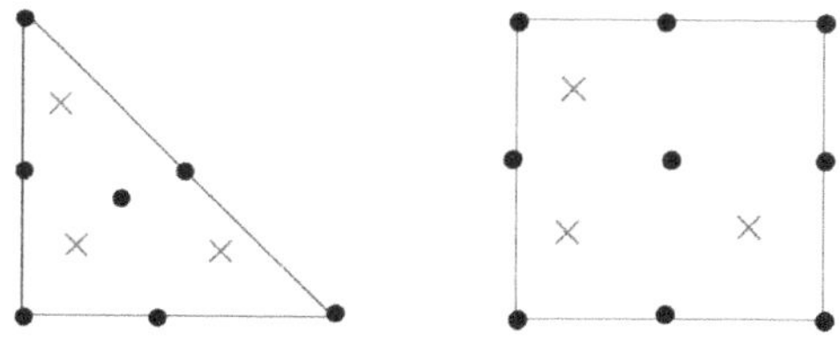

(a) $\mathbb{P}_2-$bubble$/\mathbb{P}_1$ (b) $\mathbb{Q}_2/\mathbb{P}_1$

Figure 4.5. *Crouzeix–Raviart elements: • velocity field nodes, × pressure field nodes. For a color version of this figure, see www.iste.co.uk/elhami/interactions.zip*

4.4. Finite element method for the solid

We wish to solve the following equation on the structure domain. If there are no volume sources, the equation describing its vibrational behavior is given by:

$$\rho_s \frac{\partial^2 u}{\partial t^2} - \nabla \cdot \sigma(u) = 0 \quad \text{on} \quad \Omega_S \qquad [4.51]$$

together with its boundary and coupling conditions. In our case, the material chosen for the load-bearing edge is a homogeneous and isotropic plastic. The stress–strain relation is given by the generalized Hooke's law:

$$\sigma = D\varepsilon \qquad [4.52]$$

where $\sigma = \sigma_{xx}\sigma_{yy}\sigma_{zz}\sigma_{xy}\sigma_{yz}\sigma_{zx}$ is the stress vector, ε is the strain vector and D is the stiffness tensor. In the case of a structure consisting of a homogeneous and isotropic material, D is calculated from the Lamé coefficients λ and μ, the Young's E modulus and the Poisson coefficient ν:

$$\lambda = \frac{E\nu}{(1+\nu)(1-2\nu)} \quad \mu = \frac{E}{2(1+\nu)} \qquad [4.53]$$

$$D = \begin{bmatrix} \lambda+2\mu & \lambda & \lambda & 0 & 0 & 0 \\ \lambda & \lambda+2\mu & \lambda & 0 & 0 & \\ \lambda & \lambda & \lambda+2\mu & 0 & 0 & 0 \\ 0 & 0 & 0 & \mu & 0 & 0 \\ 0 & 0 & 0 & 0 & \mu & 0 \\ 0 & 0 & 0 & 0 & 0 & \mu \end{bmatrix} \quad [4.54]$$

The strain-displacement relation is nonlinear:

$$\varepsilon = \frac{1}{2}\left(\nabla u +^t \nabla u\right) + \frac{1}{2}\left({}^t\nabla u \nabla u\right) \quad [4.55]$$

4.4.1. *Discretization*

Discretizing with the finite element method requires a weak formulation (see above). First, we subdivide the domain Ω_s into multiple finite elements (Ω_e). This mesh must be constructed from reference elements [DHA 04]. For each finite element, we define a set of points called nodes, to which we assign degrees of freedom (nodal displacement). We now seek to define approximate solutions for each of these elements.

We calculate the displacement field u on each finite element Ω_e from a finite number of values at given points (nodes) on the finite element by polynomial interpolation. We therefore obtain a linear equation of the following type on the element Ω_e:

$$U(x, y, z, t) = N_e(x, y, z)U_e(t) \quad [4.56]$$

where

– U_e is the vector of nodal unknowns for the displacement on the finite element Ω_e;

– N_e is the matrix of shape functions (interpolation polynomials) for the finite element Ω_e.

The scalar product of the two vectors uv may also be written as ${}^tvu$. The same is true for the contraction of the two tensors $\sigma\varepsilon$, which can be written as ${}^t\varepsilon\sigma$.

To discretize the integral function, we will usually choose the Galerkin method, which consists of choosing the same approximation functions for the virtual displacement field as for the physical displacement field (nodal displacement). This method has the advantage of resulting in symmetric systems of equations. The obtained equation thus becomes:

$$ {}^t\Phi_e m_e \ddot{u}_e + {}^t\Phi_e k_{\varepsilon_e} u_e + {}^t\Phi_e k_{\omega_e}(u) u_e - {}^t\Phi_e f_e = 0 \qquad [4.57] $$

where the elementary mass matrices m_e, the linear part of the stiffness k_{ε_e}, the nonlinear part of the stiffness $k_{\omega_e}(u)$ and the vector of external forces f_e are defined by:

$$ m_e = \int_{\Omega_e} \rho_S {}^t N_e N_e d\Omega_e $$

$$ k_{\varepsilon_e} = \int_{\Omega_e} {}^t B_e D B_e d\Omega_e $$

$$ k_{\omega_e} = \int_{\Omega_e} {}^t G_e D u_e^t \, {}^t G_e G_e d\Omega_e $$

$$ f_e = \int_{\Gamma_e} {}^t N_e \sigma(u) n d\Gamma_e $$

and the matrices of derivatives of the shape functions are:

$$ B = \begin{bmatrix} \frac{\partial N_1}{\partial x_1} & 0 & 0 \\ 0 & \frac{\partial N_2}{\partial x_2} & 0 \\ 0 & 0 & \frac{\partial N_3}{\partial x_3} \\ \frac{1}{2}\frac{\partial N_1}{\partial x_2} & \frac{1}{2}\frac{\partial N_2}{\partial x_1} & 0 \\ 0 & \frac{1}{2}\frac{\partial N_2}{\partial x_3} & \frac{1}{2}\frac{\partial N_3}{\partial x_2} \\ \frac{1}{2}\frac{\partial N_1}{\partial x_3} & 0 & \frac{1}{2}\frac{\partial N_3}{\partial x_1} \end{bmatrix} \qquad [4.58] $$

$$ G = \frac{1}{2} \begin{bmatrix} \frac{\partial N_1}{\partial x_1} & \frac{\partial N_1}{\partial x_2} & \frac{\partial N_1}{\partial x_3} \\ \frac{\partial N_2}{\partial x_1} & \frac{\partial N_2}{\partial x_2} & \frac{\partial N_2}{\partial x_3} \\ \frac{\partial N_3}{\partial x_1} & \frac{\partial N_3}{\partial x_2} & \frac{\partial N_3}{\partial x_3} \end{bmatrix} \qquad [4.59] $$

4.4.2. *Assembling the system*

The system is constructed by assembling the equations of the different elements. After eliminating the trivial solution $\Phi = 0$, we arrive at an equation of the form:

$$M\ddot{u} + K_\varepsilon u + K_\omega(u)u - F = 0 \qquad [4.60]$$

The boundary conditions allow us to state that the vector of external forces is zero on the free boundaries. In our case, the boundaries of the domain are either fixed or in contact with the fluid. In the latter case, the external forces are equal to the forces applied by the fluid on the structure. This system requires special solving methods due to the nonlinearity of the stiffness.

4.4.3. *Solving the system of algebraic equations*

To solve the system, we need to be able to calculate the integrals in the mass and stiffness matrices, and advance the time of the problem while taking into account its nonlinearity.

4.4.4. *Integration by Gaussian quadrature*

The reference elements allow us to perform integration over the range $[-1, +1]$ using the method of Gaussian quadrature. The surface and volume forces are stated below:

$$\int_{-1}^{1}\int_{-1}^{1} \psi(x,y)dxdy \approx \sum_{i=1}^{i=n}\sum_{j=1}^{j=m} \omega_i\omega_j\psi(x_i,y_j) \qquad [4.61]$$

$$\int_{-1}^{1}\int_{-1}^{1}\int_{-1}^{1} \psi(x,y,z)dxdydz \approx \sum_{i=1}^{i=n}\sum_{j=1}^{j=m}\sum_{k=1}^{k=l} \omega_i\omega_j\omega_k\psi(x_i,y_j,z_k) \qquad [4.62]$$

where ψ is the function that we wish to integrate, the ω_i are the weights and the x_i are the coordinates of the integration points (see Figure 4.6).

This method offers an excellent level of precision, since it calculates exactly the integrals of polynomials of order $2N - 1$.

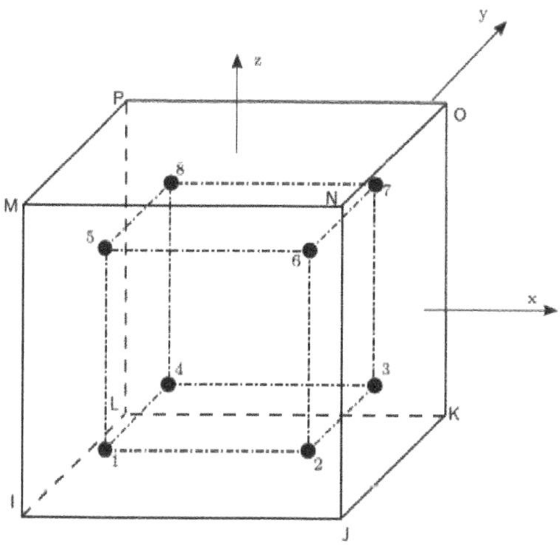

Figure 4.6. *Positions of the Gaussian points 1–8 for a quadrangular element*

4.4.5. *Advancing the time step using the Hilbert–Hugues–Taylor algorithm*

To advance time, we use an implicit algorithm called Hilber–Hugues–Taylor [ANS 15], which is accurate to the second order and unconditionally stable (with the given choice of parameters). This algorithm is a more general version of Newmark's algorithm:

$$M\ddot{u}_{n+1-\alpha_m} + C\dot{u}_{n+1-\alpha_f} + Ku_{n+1-\alpha_f} = F_{n+1-\alpha_f} \qquad [4.63]$$

with the following subrelaxed forms:

$$\ddot{u}_{n+1-\alpha_m} = (1-\alpha_m)\ddot{u}_{n+1} + \alpha_m\ddot{u}_n \qquad [4.64]$$

$$\dot{u}_{n+1-\alpha_f} = (1-\alpha_f)\dot{u}_{n+1} + \alpha_f\dot{u}_n \qquad [4.65]$$

$$u_{n+1-\alpha_f} = (1-\alpha_f)u_{n+1} + \alpha_f u_n \qquad [4.66]$$

$$F_{n+1-\alpha_f} = (1-\alpha_f)F_{n+1} + \alpha_f F_n \qquad [4.67]$$

Approximating the time advancement by finite differences, we can show that:

$$\ddot{u}_{n+1} = a_0(u_{n+1} - u_n) - a_2\dot{u}_n - a_3\ddot{u}_n \quad [4.68]$$

$$\dot{u}_{n+1} = \dot{u}_n + [(1-\delta)\ddot{u}_n + \ddot{u}_{n+1}\delta]\Delta t \quad [4.69]$$

and

$$u_{n+1} = u_n + \dot{u}_n\Delta t + [(\frac{1}{2} - \alpha)\ddot{u}_n + \alpha\ddot{u}_{n+1}]\Delta t^2 \quad [4.70]$$

where Δt is the time step between n and $n+1$, and δ and α are the Newmark integration parameters. Substituting equations [4.64]–[4.67] into equation [4.63], we obtain the following expression for the vector of unknowns at the time step $n+1$:

$$\begin{aligned}&(a_0M + a_1C + (1-\alpha_f)K)u_{n+1}\\&= (1-\alpha_f)F_{n+1} + \alpha F_n + M(a_0u_n + a_2\dot{u}_n + a_3\ddot{n})\\&\qquad + C(a_1u_n + a_4\dot{u}_n + a_5\ddot{n}) - K\alpha_f u_n\end{aligned} \quad [4.71]$$

with the following parameters:

$$\begin{aligned}a_0 &= \frac{1-\alpha_m}{\alpha\Delta t^2} & a_1 &= \frac{(1-\alpha_f)\delta}{\alpha\Delta t}\\ a_2 &= \frac{1-\alpha_m}{\alpha\Delta t} & a_3 &= \frac{1-\alpha_m}{2\alpha} - 1\\ a_4 &= \frac{(1-\alpha_f)\delta}{\alpha} - 1 & a_5 &= (1-\alpha_f)\left(\frac{\delta}{2\alpha} - 1\right)\Delta t\end{aligned}$$

and

$$\alpha = \frac{1}{4}(1+\gamma)^2 \qquad \delta = \frac{1}{2} + \gamma \qquad \alpha_f = \gamma \qquad \alpha_m = 0 \quad [4.72]$$

where γ is the amplitude reduction factor, taken to have value 0.005.

Equation [4.71] can be written as the product of a coefficient matrix A and the vector of unknowns at time $n + 1$:

$$A(u_{n+1})u_{n+1} = F(u_{n+1}) \quad [4.73]$$

4.4.6. *Linearization using the Newton–Raphson algorithm*

The Newton–Raphson method can be used to solve this nonlinear system. The solution vector at time $n + 1$ is the sum of the solution vector at time n and an increment Δu_n:

$$u_{n+1} = u_n + \Delta u_n \quad [4.74]$$

The increment of the solution vector is found by solving:

$$J(u_n)\Delta u_n = -R(u_n) \quad [4.75]$$

where $R = A(u_n)u_n - F(u_n)$ and $J(u_n)$ is the tangent matrix of R at time n:

$$J(u_n) = \nabla R \quad [4.76]$$

We obtain a converged solution when:

$$\|R(u_n) - R(u_{n+1})\| < \varepsilon_R R_{Ref} \quad [4.77]$$

where $\|R\|$ is the Euclidean norm given by:

$$\|R\| = \left(\sum_i R_i^2\right)^{\frac{1}{2}} \quad [4.78]$$

and the default values are $R_{Ref} = 0.01$ and $\varepsilon_R = 0.001$, corresponding to a convergence criterion of 10^{-5}. At each iteration of the Newton–Raphson algorithm, we must solve a linear system of the form $Ku = F$. To do this, we use the method known as the preconditioned conjugate gradient method.

4.5. Finite volumes for the fluid

4.5.1. *Generic transport equation*

In numerical analysis, the finite volume method is used to numerically solve partially differential equations, similarly to the finite element and finite difference methods. Unlike the finite difference method, which involves approximations of the derivatives, the finite volume and finite element methods use approximations of the integrals and apply Gauss's divergence theorem. However, the finite volume method is directly based upon the so-called strong form of the equation to be solved, whereas the finite element method is based on a variational formulation of the equation.

The technique of control volumes integrates the partial differential equations over each control volume to find discretized equations that conserve all physical quantities over each control volume.

The conservation equations have the same form, which we can reformulate into the general form used to represent the transport equation of a scalar property, with the notation popularized by Bird *et al.* [BIR 07] and later reused by Brodkey and Hershey, two of the biggest names in transport phenomena. Specifically, we use the expression:

$$\frac{\partial \rho\phi}{\partial t} + \nabla \cdot J = S^{\phi} \qquad [4.79]$$

where ϕ is a scalar representing properties, J denotes the convection flux and the diffusion flux of ϕ, defined as $J = \rho\phi u \Gamma_{\phi} \nabla\phi$, Γ_{ϕ} is the diffusion coefficient of ϕ and S^{ϕ} is the source term associated with the variable ϕ.

This convection–diffusion equation, which may also be written in the form:

$$\frac{\partial \rho\phi}{\partial t} + \mathrm{div}(\rho\phi u) = \mathrm{div}(\Gamma_{\phi} \nabla\phi) + S^{\phi} \qquad [4.80]$$

intrinsically contains representations of several transport equations in the dependent variable ϕ. In general, each term has a well-defined physical interpretation.

For example if $\phi = 1$ and $S^{\phi} = 0$, we recover the continuity equation:

$$\frac{\partial \rho}{\partial t} + \mathrm{div}(\rho u) = 0 \qquad [4.81]$$

If $\phi = T$, $S^{\phi} = Q(T)$ and $\Gamma_{\phi} = k/c_p$, we recover a simplified form of the energy equation:

$$\frac{\partial \rho T}{\partial t} + \text{div}(\rho u T) = \text{div}\left(\frac{K}{c_p}\nabla T\right) + Q(T) \qquad [4.82]$$

and we can similarly recover most equations describing the transport of a quantity. The generalized transport equation is written in the form of a divergence, which allows us to apply Gauss's theorem when working with the integral equations.

4.5.2. *Conservation property of the method*

The fundamental property of the finite volume method is conservation, which follows from the integral form of the method. This means that when we consider the integral form of the general transport equation:

$$\frac{\partial}{\partial t}\int_{\Omega} \rho\phi d\Omega + \oint_{\Gamma} J \cdot n d\Gamma = \int_{\Omega} S^{\phi} d\Omega \qquad [4.83]$$

Note that the variation of the property ϕ depends on the net flux through the surface Γ enclosing the volume Ω. This is a crucial observation, since if we partition the domain into multiple subdomains, each with an arbitrary number of sides, we obtain the following result:

$$\int_{A}^{B} J \cdot n d\Gamma = -\int_{B}^{A} J \cdot n d\Gamma \qquad [4.84]$$

where *A* and *B* are the two edges of the side.

In other words, independent of the choice of partition, the internal flux always cancels, guaranteeing that the global property ϕ is conserved.

4.5.3. *The different steps in the method*

To solve the generalized transport equation with a finite volume method, we must apply the volume and contour integrals to a set of elements, and then replace these integrals with algebraic expressions. Thus, we begin by partitioning the domain into a finite set of subdomains, and then we calculate the value of the ϕ variables in each element or control volume. The steps of the method can be summarized as follows:

– Partition the region of interest into subdomains or control volumes. The control volumes must completely cover the domain, and they can have different shapes.

The most common basis elements are quadrilaterals and triangles in two dimensions, and hexahedra and tetrahedra in three dimensions.

– Integrate the equations over each volume and apply Gauss's divergence theorem. On each volume, we replace the function ϕ with an approximation; generally a constant associated with the center of gravity of the element. We also estimate the integral of the flux over the sides of the control volume, usually by taking the average value.

– Incorporate the boundary conditions.

– Find the sum of ϕ on each volume.

– Solve the resulting algebraic system.

4.5.4. *Integrating the model equation*

We integrate the generalized transport equation over the time interval Δt and the control volume Ω:

$$\int_t^{t+\Delta t}\int_\Omega \frac{\partial \rho\phi}{\partial t} d\Omega dt + \int_t^{t+\Delta t}\int_\Omega \nabla \cdot J d\Omega dt = \int_t^{t+\Delta t}\int_\Omega S^\phi d\Omega dt \quad [4.85]$$

The form $\nabla \cdot J$ prompts us to apply Gauss's theorem, thus:

$$\int_t^{t+\Delta t}\int_\Omega \frac{\partial \rho\phi}{\partial t} d\Omega dt + \int_t^{t+\Delta t}\int_\Gamma J \cdot \hat{n} d\Gamma dt = \int_t^{t+\Delta t}\int_\Omega S^\phi d\Omega dt \quad [4.86]$$

Approximating with respect to time:

The variable ϕ can be integrated with respect to time using a numerical procedure that yields the field of the variable ϕ at the end of each time step. Although the integral of the time derivative is direct, we introduce a numerical approximation that leads to the same result. This allows us to more easily conceptualize the spatial terms as an Eulerian approximation, but there also exist other methods such as Runge-Kutta and Adams. We therefore have:

$$\int_{\Omega}\int_{t}^{t+\Delta t}\frac{(\rho\phi)^{n+1}-(\rho\phi)^{n}}{\Delta t}dtd\Omega=\int_{\Omega}[(\rho\phi)^{n+1}-(\rho\phi)^{n}]d\Omega \quad [4.87]$$

and

$$\int_{\Omega}[(\rho\phi)^{n+1}-(\rho\phi)^{n}]d\Omega=-\int_{t}^{t+\Delta t}\int_{\Gamma}J\cdot\hat{n}d\Gamma dt+\int_{t}^{t+\Delta t}\int_{\Omega}S^{\phi}d\Omega dt \quad [4.88]$$

Approximating with respect to space:

By far the most common method of approximation is to assume that a property is constant on a given control volume at a given time. This property is imagined to be attached to the center of gravity, which is typically denoted by the index "P". Similarly, we can assume that the values of ϕ, u and $\nabla\phi$ are constant on each of the faces of the control volume. But piecewise constant variables are not our only option, as properties can also be allowed to vary linearly, quadratically or at higher orders within the elements as well as on their faces. In other words, the integral form inherently contained in the finite volume method allows more than one node to be incorporated into the calculations.

4.5.5. *Control volumes*

Since the finite volume method calculates the net value of a quantity on a polygon or a polyhedron, it is clear that this method is applicable to any kind of mesh. In this sense, it is as flexible as the finite element method when considering general geometries.

From a historical perspective, the finite volume method was originally applied to meshes with structures that were logically ordered along the coordinate axes. In this case, we speak of a structured mesh with a logical organization of type "finite differences". However, this is not necessarily the case, and a mixed mesh composed of triangles and quadrilaterals can also be used.

Regardless of the choice of discretization, we must determine the positions at which to calculate the variables. If we assume that the computation domain is subdivided by a primary mesh, it is often easy to identify a number of natural choices for the position relative to this mesh. In particular, we can identify:

– a point inside the polygon, typically its center of gravity. With this discretization, the control volume is the same as the initial mesh, and geometric reconstruction is not required (see Figure 4.7(a));

– a vertex of the polygon. The computational element is constructed by dividing each primary element of the mesh into three or four subdivisions for triangles and quadrilaterals, respectively. All of these subelements are then assembled around their shared node. With this choice, reconstructing the control volumes results in what is known as the dual mesh (see Figure 4.7(b));

– the center of a face. In this case also, the primary elements of the mesh are first subdivided before forming the control volumes by connecting together subelements that share an edge. As before, this gives rise to a dual mesh (see Figure 4.7(c)).

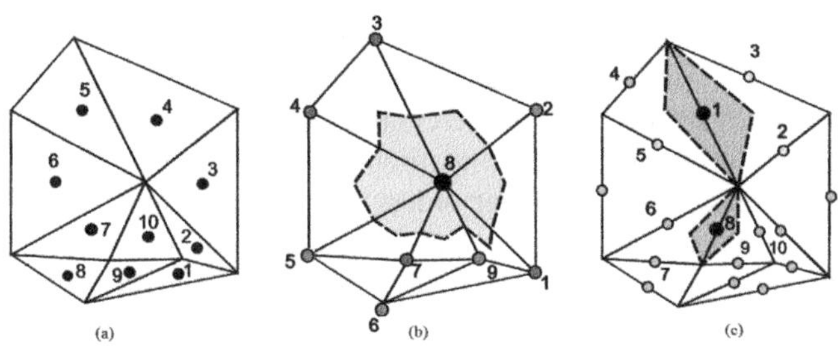

Figure 4.7. *a) Center-stored control volume, b) vertex-stored control volume and c) face-stored control volume*

Although the geometric shapes of the control volumes obtained from each choice are of course different, they simply represent alternative ways of defining the space associated with the averaged value of the considered property. In the first case, the control volume is the polygon itself, whereas in the other two cases the control volume is formed by assembling parts from several different polygons. These three options are described using "center-stored", "vertex-stored" and "face-stored" variables . However, it is important to note that these terms reference the primary mesh. We should always bear in mind that with respect to the control volume itself, regardless of how it was created, the properties are always constant on each element and are associated with its center.

Now that we have identified the different ways of storing variables on each control volume, we must find the flux through the faces in order to calculate an average value for each property.

Since the flux does not just have a geometry but also has a neighborhood of properties, we must define a new medium in addition to the geometric medium. Indeed, if by definition we assumed that a property was constant on the control volume, then it must necessarily be unknown, and would typically be discontinuous at the interface.

4.5.6. *Physical interpolation*

When interpolating a quantity transported by a moving fluid, as well as considering the position at which it is defined, we must take into account the nature of the flow. Several schemes have been proposed to push back these limitations. The difference between their approaches essentially lies in the choice of strategy for estimating the convection flux, which plays a crucial role in the stability of the numerical scheme.

In the following section, we will review a few basic schemes using the notation introduced by Imperial College and widely popularized by Patankar [PAT 80].

Since the property averages are always calculated in connection with so-called center-stored control volumes, we will restrict attention to aspects specific to this type of control volume. In general, we assume that the properties are stored at the center of gravity, which we denote by the index P.

4.5.7. *Evaluating the flux through the faces*

To illustrate the difficulties associated with evaluating the flux through the interface, we will study the 1D *linear* equation with a generic property ϕ, representing a steady problem. This is given by:

$$\frac{d}{dx}(\rho u \phi) - \frac{d}{dx}\left(\Gamma \frac{d\phi}{dx}\right) = 0 \qquad [4.89]$$

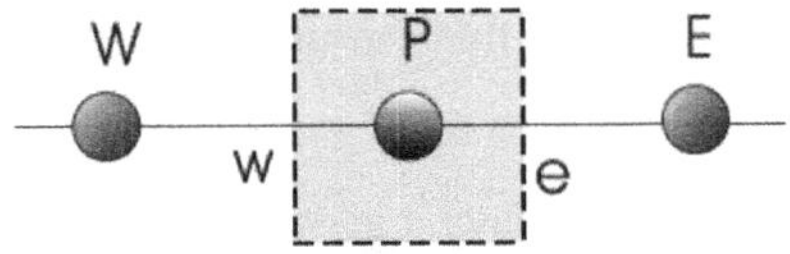

Figure 4.8. *Control voluume*

4.5.8. *Centered scheme*

Given a uniform "mesh" $\Delta x = cte$, the value of the variable ϕ at the interface w is approximated by:

$$\phi_w = \frac{\phi_W + \phi_P}{2} \qquad \left.\frac{d\phi}{dx}\right|_w = \frac{\phi_P - \phi_W}{\Delta x} \qquad [4.90]$$

Now, the flux $J = (\rho u \phi) - \left(\Gamma \frac{\partial \phi}{\partial x}\right)$ through the faces w and e is given by:

$$J_w = \frac{\phi_P + \phi_W}{2} \rho u_w - \Gamma_w \frac{(\phi_P - \phi_W)}{\Delta x}$$
$$J_e = \frac{\phi_P + \phi_W}{2} \rho u_e - \Gamma_e \frac{(\phi_E - \phi_P)}{\Delta x}$$

Defining $J = (\rho u \phi) - \Gamma \nabla \phi$, the equation to be solved is given by $\frac{d}{dx}(J) = 0$, the discrete form of which is: $J_e - J_w = 0$. Therefore, with this centered scheme, we find:

$$\frac{1}{2}\rho u_e(\phi_P + \phi_E) - \frac{1}{2}\rho u_w(\phi_P + \phi_W) = \frac{\Gamma_e}{\Delta x}(\phi_E - \phi_P) - \frac{\Gamma_w}{\Delta x}(\phi_P - \phi_W) \qquad [4.91]$$

Using the notation $F = \rho u$ and $D = \frac{\Gamma}{\Delta x}$, for the convection and diffusion flux, respectively, we can write

$$\frac{1}{2}F_e(\phi_P + \phi_E) - \frac{1}{2}F_w(\phi_P + \phi_W) = D_e(\phi_E - \phi_P) - D_w(\phi_P - \phi_W) \qquad [4.92]$$

and therefore:

$$\left[D_e - \frac{1}{2}F_e\right]\phi_E + \left[D_w - \frac{1}{2}F_w\right]\phi_W = \left[\frac{1}{2}F_e - \frac{1}{2}F_w + D_e + D_w\right]\phi_P \qquad [4.93]$$

With the following definitions for the coefficients of ϕ_E, ϕ_W and ϕ_P:

$$a_E = \left[D_e - \frac{1}{2}F_e\right], \qquad a_W = \left[D_w + \frac{1}{2}F_w\right]$$

and

$$a_P = \left[\frac{1}{2}F_e - \frac{1}{2}F_w + D_e + D_w\right]$$

We can represent this 1D convection–diffusion problem by the equation:

$$a_P\phi_P = a_E\phi_E + a_W\phi_W$$

For incompressible flows, the flow field must have zero divergence:

$$\nabla \cdot u = 0 \qquad [4.94]$$

In our case, this condition translates to:

$$F_e = F_w$$

and allows us to simplify the equations as a result of the relation between the coefficient of ϕ_P and the coefficients of the neighboring variables ϕ_W and ϕ_E

$$a_P = \left[\frac{1}{2}F_e - \frac{1}{2}F_w + D_e + D_w\right] = a_E + a_W + F_e - F_w \qquad [4.95]$$

Thus, once the continuity equation has been satisfied, we find the following relation between these coefficients:

$$a_P = a_E + a_W \qquad [4.96]$$

4.5.9. *Upwind scheme*

This scheme proposes to define the value of the convected property at the interfaces of the control volumes as a function of the direction of convection, which means that it depends on the direction of the flow:

$$\phi_e = \begin{cases} \phi_E & \text{if} \quad F_e < 0 \\ \phi_P & \text{if} \quad F_e > 0 \end{cases} \qquad [4.97]$$

The discrete equation corresponding to the 1D convection–diffusion equation, $\frac{d}{dx}(\rho u \phi) = \frac{d}{dx}\left(\Gamma \frac{d\phi}{dx}\right)$, may be written as:

$$\frac{\rho u|_e \phi_e - \rho u|_w \phi_w}{\Delta x} = \frac{\Gamma \frac{\phi_E - \phi_P}{\Delta x} - \Gamma \frac{\phi_P - \phi_W}{\Delta x}}{\Delta x} \qquad [4.98]$$

We now consider the Navier–Stokes equations. These equations include terms for the convection and the diffusion, as well as source terms. They can be obtained from the generalized transport equation:

$$\frac{\partial(\rho\phi)}{\partial t} + \nabla \cdot J = \hat{S} \qquad [4.99]$$

where $J = \rho u \phi - \Gamma \vec{\nabla} \phi$. By initially setting $\phi = 1$, we obtain the continuity equation:

$$\frac{\partial \rho}{\partial t} + \nabla \cdot (\rho u) = 0 \qquad [4.100]$$

Now, setting $\phi = u$, $\Gamma = \mu$ and $S = \nabla P$, we recover the equation of motion:

$$\frac{\partial \rho u}{\partial t} + \nabla \cdot (\rho u u) = \nabla \mu u - \nabla P \qquad [4.101]$$

4.5.10. *Hybrid scheme*

This scheme arises as a combination of the centered scheme and the upwind scheme, allowing either one to be used depending on the value of the local Péclet number, which describes the ratio between the convective forces and the diffusive forces:

$$Pe = \frac{F}{D} = \frac{\rho u \Delta x}{\Gamma_\phi} \qquad [4.102]$$

– if $Pe \leq 2$, the centered scheme is used;

– if $Pe > 2$, the upwind scheme is used.

4.5.11. *Discretization*

We will present the discretization procedure in the context of a 2D problem using the equation of motion in the x direction written in dimensionless form. The integral form of this equation can be stated as:

$$\int \frac{\partial u}{\partial t} dV + \int uudy + \int uvdx = \frac{1}{Re}\left[\int \frac{\partial u}{\partial x} dy + \int \frac{\partial u}{\partial y} dx\right] - \int Pdy \qquad [4.103]$$

After discretizing by finite volumes, equation [4.103] may be written as:

$$\frac{(\rho u)_p^{n+1} - (\rho u)_p^n}{\Delta t}\Delta V + [\rho U_e u_e - \rho U_w u_w]\Delta y + [\rho U_n u_n - \rho U_s u_s]\Delta x$$
$$= \frac{1}{Re}\left.\frac{\partial u}{\partial x}\right|_e \Delta y - \frac{1}{Re}\left.\frac{\partial u}{\partial x}\right|_w \Delta y$$
$$+ \frac{1}{Re}\left.\frac{\partial u}{\partial y}\right|_n \Delta x \quad \frac{1}{Re}\left.\frac{\partial u}{\partial y}\right|_s \Delta x - (P_e - P_w)\Delta y$$

where the symbols U and u were used to, respectively, denote the normal speed (mass flow rate) and the convected property.

Diffusion terms

In all schemes, the natural discretization of the diffusion terms is given by a centered calculation. The various terms are approximated by:

$$\left(\frac{\partial u}{\partial x}\right)_e \simeq \frac{u_E - u_P}{\Delta x} \qquad \left(\frac{\partial u}{\partial x}\right)_w \simeq \frac{u_P - u_W}{\Delta x}$$

$$\left(\frac{\partial u}{\partial y}\right)_s \simeq \frac{u_P - u_S}{\Delta y} \qquad \left(\frac{\partial u}{\partial y}\right)_n \simeq \frac{u_N - u_P}{\Delta y}$$

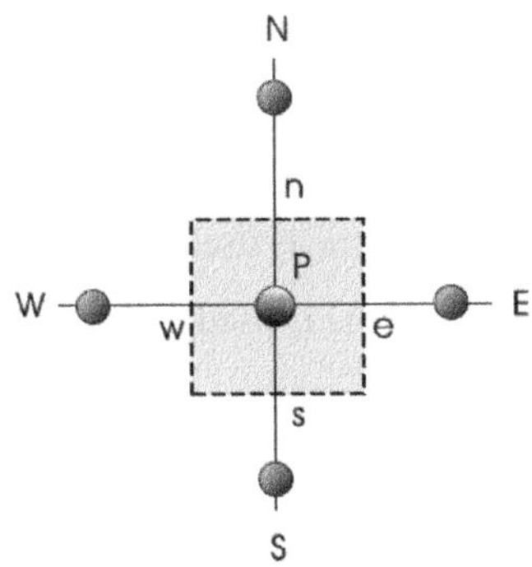

Figure 4.9. *Notation for storing variables*

Convection terms

The convection terms at each of the faces are approximated by performing an interpolation that considers the relative significance of the convective and diffusive terms.

Mass flow rate

The mass per unit time and per unit area at the interfaces (symbol U) is estimated by calculating a simple arithmetic average. In 2D, if the velocity components are stored at the same position, we have:

$$U_w = \frac{u_W + u_P}{2} \qquad U_e = \frac{u_P + u_E}{2}$$

$$U_n = \frac{v_N + v_P}{2} \qquad U_s = \frac{u_P + u_S}{2}$$

Discrete equation

If we substitute the expressions for the diffusion and convection terms into equation [4.106], we find:

$$(u_P^{n+1} - u_P^n)\frac{\Delta V}{\Delta t} = \left[(\frac{1}{2} + \alpha_w)U_w\Delta y + \frac{1}{Re}\frac{\Delta y}{\Delta x}\right] u_W$$

$$
\begin{aligned}
&+\left[-(\frac{1}{2}+\alpha_e)U_e\Delta y+\frac{1}{Re}\frac{\Delta y}{\Delta x}\right]u_E\\
&+\left[(\frac{1}{2}+\alpha_s)U_s\Delta x+\frac{1}{Re}\frac{\Delta x}{\Delta y}\right]u_S\\
&+\left[-(\frac{1}{2}+\alpha_n)U_n\Delta x+\frac{1}{Re}\frac{\Delta x}{\Delta y}\right]u_N\\
&+\left[(\frac{1}{2}-\alpha_w)U_w\Delta y-(\frac{1}{2}+\alpha_e)U_e\Delta y+(\frac{1}{2}-\alpha_s)U_s\Delta x\right.\\
&\left.-(\frac{1}{2}+\alpha_n)U_n\Delta x-\frac{2}{Re}\frac{\Delta y}{\Delta x}-\frac{2}{Re}\frac{\Delta x}{\Delta y}\right]u_P\\
&+P_w\Delta y-P_e\Delta y
\end{aligned}
\quad [4.104]
$$

The first four coefficients of u_W, u_E, u_S and u_N are denoted by a_W, a_E, a_S and a_N, and the coefficient of u_P is denoted by $-a_P$. Note that if continuity is satisfied

$$
a_P = a_W + a_E + a_S + a_N \quad [4.105]
$$

then the discrete equation of motion can be written as:

$$
\begin{aligned}
u_P^{n+1} &= u_P^n + \frac{\Delta t}{\Delta V}[a_W u_W + a_E u_E + a_S u_S + a_N u_N - a_P u_P\\
&\quad + P_w\Delta y - P_e\Delta y]
\end{aligned}
\quad [4.106]
$$

4.6. Coupling procedures

4.6.1. *Coupling strategies*

The key step in the coupled fluid–structure computation lies in solving the numerical coupling between the fluid and structure solvers. This must be compatible with the model chosen for the physical mechanics responsible for the couplings between the two systems. As such, we frequently distinguish between different types of numerical schemes depending on whether they induce either strong or weak couplings.

The most basic procedure is fully explicit partitioned coupling, where the fluid and structure solvers alternate in exchanging information via the boundary conditions at each iteration. This technique has the advantage of being easy to implement, since the fluid and structure solvers can be freely and independently chosen. However, it does not guarantee that the coupling schemes will perform well in terms of stability.

Conversely, a fully implicit monolithic coupling procedure allows us to attempt to solve all of the unknowns of both the fluid and structure problems simultaneously in the same iterations. This approach is optimal in terms of guaranteeing convergence to the solution, but other difficulties arise: the full system can be difficult to solve, and many structural modifications will need to be made to the fluid and structure solvers. In practice, a strategy somewhere in between is chosen to combine the advantages (and avoid the disadvantages) of the two extreme approaches. One possible technique is to start with an explicit partitioned algorithm and improve its robustness and relax the limitations that it places on the time step. The idea is to make it more implicit (semiimplicit) using an iterative predictor–corrector algorithm (Figure 4.10). In order to couple the codes together, the data transfers between the fluid and solid solvers are carried out via a coupling interface. The coupling tool is a coupler that executes and monitors the data transfer at each iteration of the coupled computation. This coupler also interpolates the data between the grids of the fluid and solid solvers, if necessary.

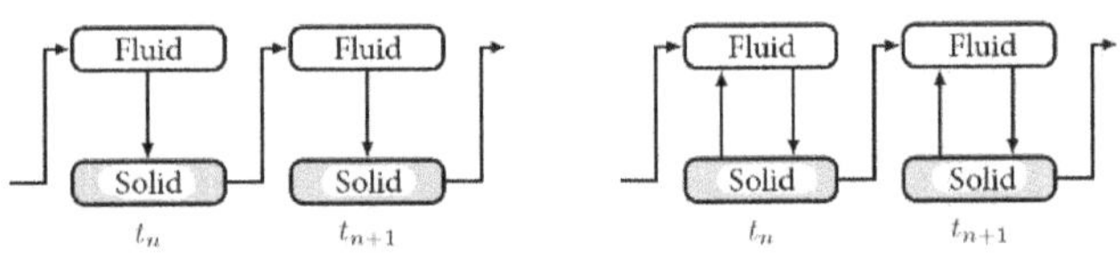

Figure 4.10. *Diagram of an explicit partitioned coupling procedure (left) and an implicit procedure with a prediction–correction algorithm (right)*

4.6.2. *Implicit partitioned coupling*

We consider an implicit partitioned coupling approach as described in [SCH 06]. A schematic overview of the iteration process is shown in Figure 4.11. After initialization, the current iteration of the flow field is determined based on the geometry of the domain. The fluid loads on the walls are deduced and sent to the structure solver via the boundary conditions. The structure solver then calculates the displacement of the structure, which will be used to update the fluid domain and its mesh during the next iteration. The principle of the subcycling algorithm is to execute subiterations until the required level of accuracy is attained at the interface relative to the prediction.

Suppose that the fluid and structure states are given at time t^n. The kth subiteration of the $(n+1)$th time step is performed as follows:

– predict the displacement of the interface:

$$u_{ifs}^{n+1,k} = u_{ifs}(u_s^{n+1,k-1}, \dot{u}_s^{n+1,k-1}) \qquad [4.107]$$

– solve the fluid equations:

$$p^{n+1,k} = p(p^n, u_f^n, u_{ifs}^{n+1,k}) \qquad [4.108]$$

$$u_f^{n+1,k} = u(p^n, u_f^n, u_{ifs}^{n+1,k}) \qquad [4.109]$$

– calculate the fluid stresses to be applied to the structure:

$$F_f^{n+1,k} = F(p^{n+1,k}, u_f^{n+1,k}) \qquad [4.110]$$

– solve the equations relative to the structure:

$$u_s^{n+1,k} = u(u_s^n, \dot{u}_s^n, \ddot{u}_s^n, F_f^{n+1,k}) \qquad [4.111]$$

– test the convergence of the displacement:

$$\left\| \frac{u_s^{n+1,k} - u_s^{n+1,k-1}}{u_s^{n+1,0}} \right\| \leq \varepsilon \qquad [4.112]$$

– advance to the next time step if the convergence passes the test or to the next subiteration if not.

This fixed-point algorithm converges if the initial estimate is sufficiently close to the target solution. By construction, all partitioned coupling schemes are sensitive to deformations of the grid, especially during the first fluid–structure subiteration. Thus, it is possible to generate non-equilibrium solutions that can lead to instability, or which cause the iterations of the fluid–structure coupling to diverge. Adaptive subrelaxation is used to prevent this phenomenon.

By introducing a relaxation factor α_{FSI}^m, it is possible to linearly weight the calculated displacement $\tilde{u}^m$ by the value u^{m-1} of the previous iteration to deduce the new value for the displacement u^{m+1}:

$$u^{m+1} = \alpha_{FSI}^m \tilde{u}^m + (1 - \alpha_{FSI}^m) u^{m-1} \qquad [4.113]$$

where $0 < \alpha_{FSI}^m \leq 1$.

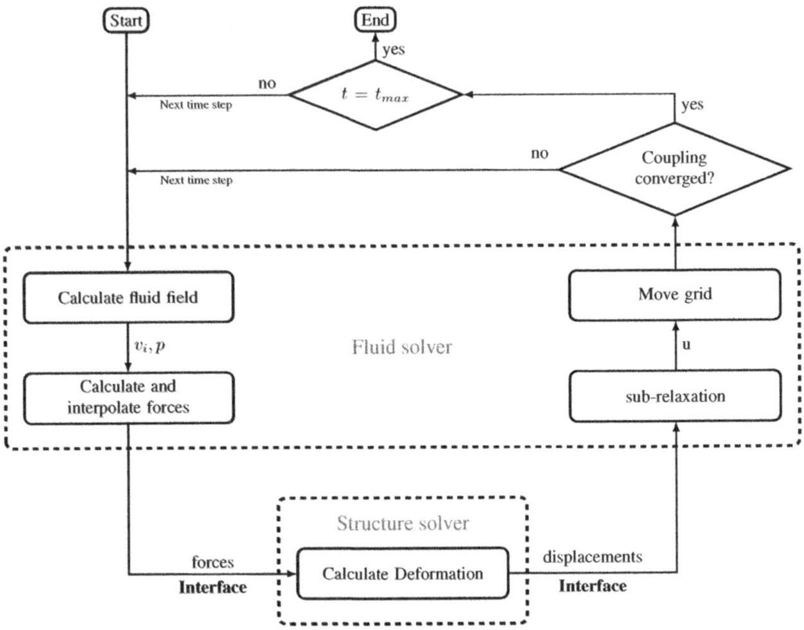

Figure 4.11. *Principle of an implicit partitioned coupling*

Subrelaxation does not alter the final solution obtained after convergence in any way [SCH 06]. There are several possible methods for determining α^m_{FSI}. One technique is to rely on an Aitken-type method (for vector equations) to accelerate linear convergence in Newton–Raphson-type iterative processes [SOU 10]. The benefit of applying this technique in the context of coupled fluid–structure simulations was shown in [SOU 10]. Using the values of the two previous iterations, the γ^m factor, called the Aitken factor, is extrapolated by:

$$\gamma^m = \gamma^{m-1} + (\gamma^{m-1} - 1)\frac{(\Delta u^{m-1} - \Delta u^m)^T \Delta u^m}{(\Delta u^{m-1} - \Delta u^m)^2} \qquad [4.114]$$

where $\Delta u^{m-1} = u^{m-2} - \tilde{u}^{m-1}$ and $\Delta u^m = u^{m-1} - \tilde{u}^m$.

The subrelaxation factor α^m_{FSI} is then given by:

$$\alpha^m_{FSI} = 1 - \gamma^m \qquad [4.115]$$

The last value of the Aitken factor in the previous iteration can be used as the first value at each time step γ^0. An arbitrary reasonable value can be chosen for the first time step.

4.7. Numerical results

4.7.1. *Static analysis*

In Ansys Workbench, the fluid–structure interaction can be simulated (one-way and two-way couplings) by connecting the coupling participants to a component called *system coupling*. A participant system is a system that sends or receives data to or from a coupled analysis. Here, Ansys/Fluent (participant 1) and Ansys/Mechanical (participant 2) are both acting in the capacity of participants [ANS 13]. In the first step, system coupling gathers information about the participants to synchronize the overall simulation, and then gives the exchanged information to each of them. The next step of the process is to organize the information exchange sequence. The solving step varies depending on the coupling method. Finally, the convergence step is executed at the end of each coupling iteration.

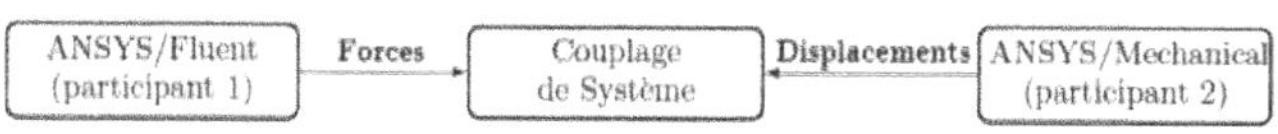

Figure 4.12. *System coupling*

Initially, geometric models of the fluid and solid domains are created with appropriate dimensions. Ansys code is used as a preprocessor to generate the geometric models. The volume mesh of the fluid domain is constructed with Gambit, and the finite element mesh is created with Ansys. The two meshes have different parameters such as the cell type, cell size and mesh resolution. The finished meshes are imported into the respective numerical solvers, where the configuration of the model simulation is implemented. The configuration of the simulation includes key steps such as the attribution of material properties, boundary conditions and numerical schemes for both models. We will perform modal analysis on the model of a 3D airplane wing. The wing has a lengthwise uniform configuration, and its cross-section is defined as a straight line and a spline. It is fixed to the body at one end, and is freely suspended at the other. The objective of this problem is to determine the natural frequencies and the deformation of the wing under the fluid flow in order to illustrate the effect of the fluid–structure interaction [ELM 15].

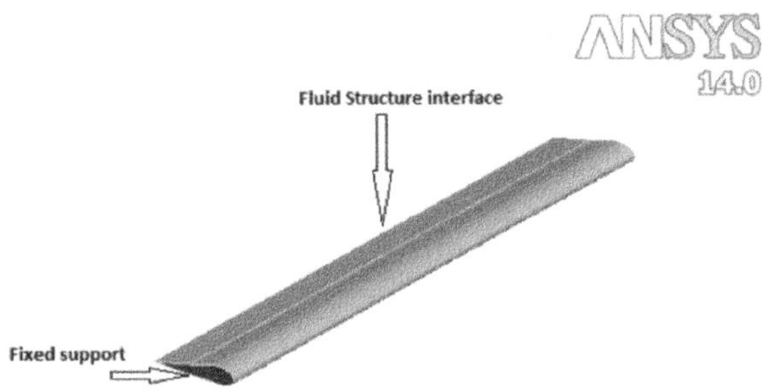

Figure 4.13. *Boundary conditions of the structure model. For a color version of this figure, see www.iste.co.uk/elhami/interactions.zip*

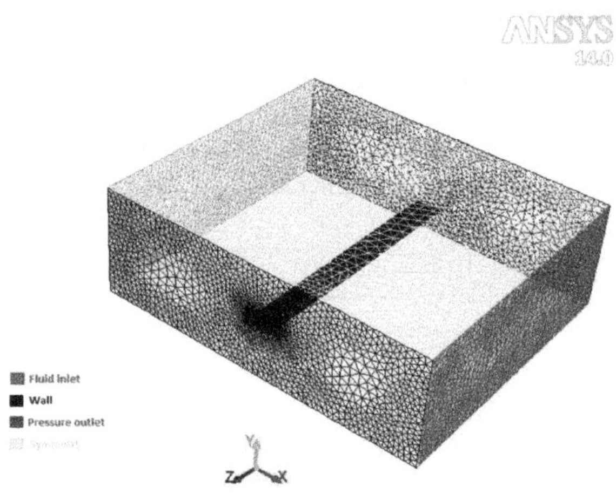

Figure 4.14. *Boundary conditions of the fluid model. For a color version of this figure, see www.iste.co.uk/elhami/interactions.zip*

In the case of a computational fluid dynamics (CFD) mesh, the surface mesh is first generated using triangular elements, which are then used to create a volume mesh. The volume mesh consists of tetrahedral cells from the category of unstructured meshes. Hexahedral cells are not used as they are not compatible with dynamic meshes in the current version of Ansys/Fluent. The mesh of the entire fluid domain is shown below in Figure 4.15.

The mesh of the structure element is created using the Ansys meshing tool as shown in Figure 4.16.

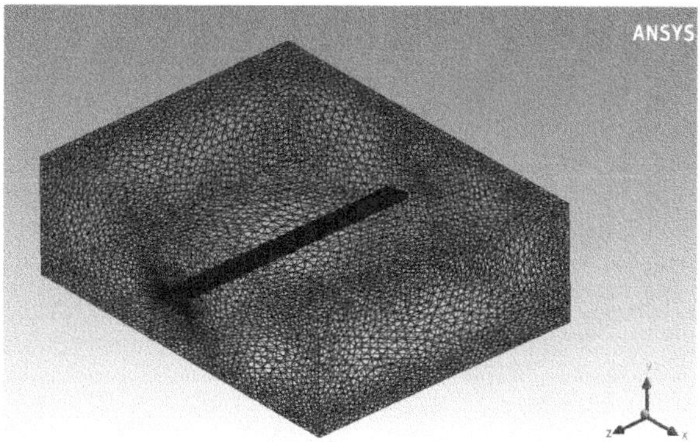

Figure 4.15. *Mesh of the fluid domain (Ansys/Fluent). For a color version of this figure, see www.iste.co.uk/elhami/interactions.zip*

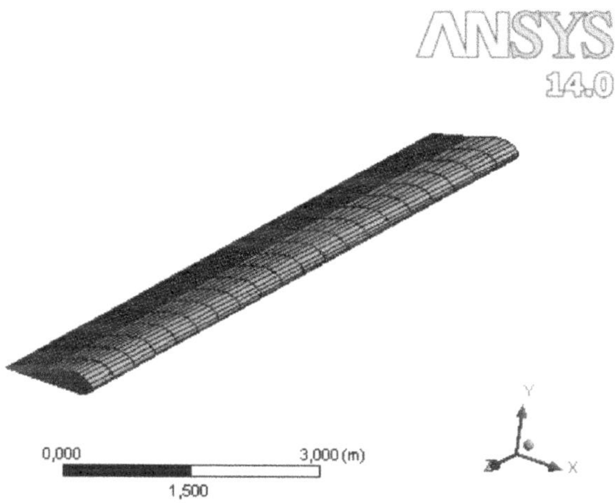

Figure 4.16. *Structure mesh. For a color version of this figure, see www.iste.co.uk/elhami/interactions.zip*

In the finite element calculation, SOLID186 is used to model the solid in three dimensions. This element is defined by 20 nodes with three degrees of freedom per node: displacement in the x, y and z directions. For the Flotran CFD element, FLUID142 is used to model the fluid flow and the interface in fluid–structure interaction problems, as shown in Figure 4.17.

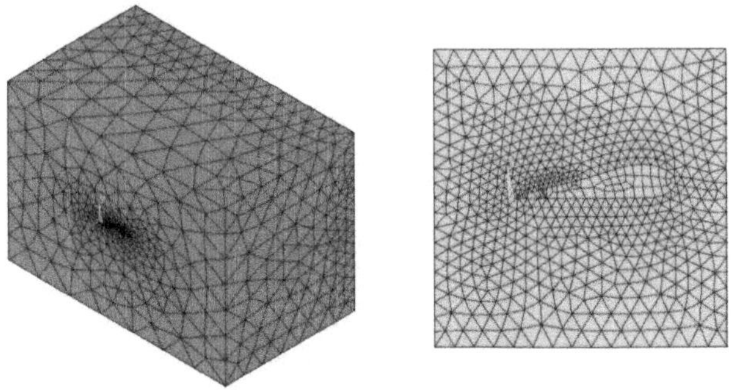

Figure 4.17. *Mesh of the fluid domain (Ansys/Flotran). For a color version of this figure, see www.iste.co.uk/elhami/interactions.zip*

As mentioned above, modal analysis using Ansys/Mechanical was performed to determine the natural frequencies and corresponding mode shapes. The analysis was performed in both of the cases presented here (Ansys/Flotran and Ansys/Fluent). First, the geometry of the deformed model subject to the flow is imported into the modal configuration of Ansys together with the boundary conditions of the fixed end, and the number of natural frequencies to be found can be chosen as an analysis parameter.

System coupling

System coupling is a tool available in Workbench to integrate different domain solvers into multiphysics simulations. The operating principle and procedure of the coupling system are described above.

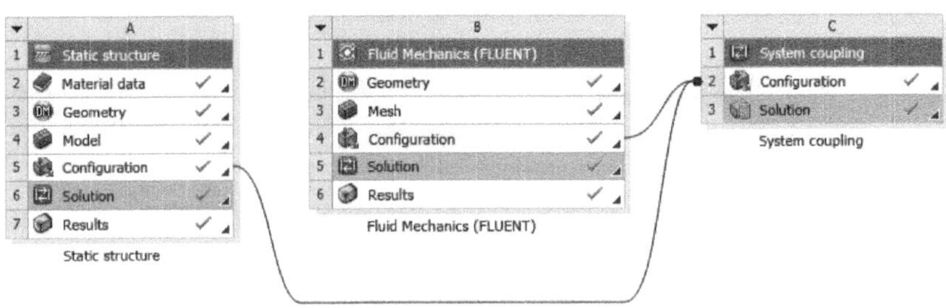

Figure 4.18. *Parameter of an FSI interface using system coupling. For a color version of this figure, see www.iste.co.uk/elhami/interactions.zip*

Figure 4.18 displays the configuration of the fluid–structure interaction using system coupling with Ansys/Fluent and Ansys/Mechanical as numerical solvers. Initially, the configurations of the simulation of the two solvers are executed, then the configuration component of the solvers is integrated into the configuration component of system coupling as shown, which allows the numerical conditions of both solvers to be synchronized and the fluid–structure interface to be identified.

The next step is to assign the simulation configurations to system coupling. It consists of three main steps:

– Analysis parameters: This parameter includes the size of the time step, the finish time and the minimum and maximum number of coupling iterations at each time step. Generally, apart from the coupling iteration at each time step, the necessary information is automatically sent to system coupling once the solvers have been coupled.

– Data transfer: This is the most vital part of the coupling system, as it incorporates and manages the data transfer sequence between the two numerical solvers. The data transfer process depends on the type of coupling. Figure 4.19 shows the data transfer for both one-way and two-way couplings. One-way coupling is performed by transferring the data in one single direction, from Ansys/Fluent to Ansys/Mechanical (forces), whereas two-way coupling involves transferring the data in both directions, i.e. first from Ansys/Fluent to Ansys/Mechanical (forces), and then from Ansys/Mechanical to Ansys/Fluent (nodal displacements).

– Simulation sequence: Here, the working sequence of the two numerical solvers must be given as an input.

In the case of Ansys/Flotran coupling, the load transfer uses physics files associating environments with each physical domain. This method is based on a single finite element mesh shared by all environments, requiring physics files to be created to define the environment of each domain; these files configure the database and prepare the shared mesh for each physics simulation. The general procedure is to read the first physics environment, and then solve it. Then, read the next environment, specify the loads to be transferred, and solve the second physics environment. The LDREAD command allows different physics environments to be connected together, and enables the data of the results specified by the first physics environment to be applied as loads in the solution of the next environment through a similar interface.

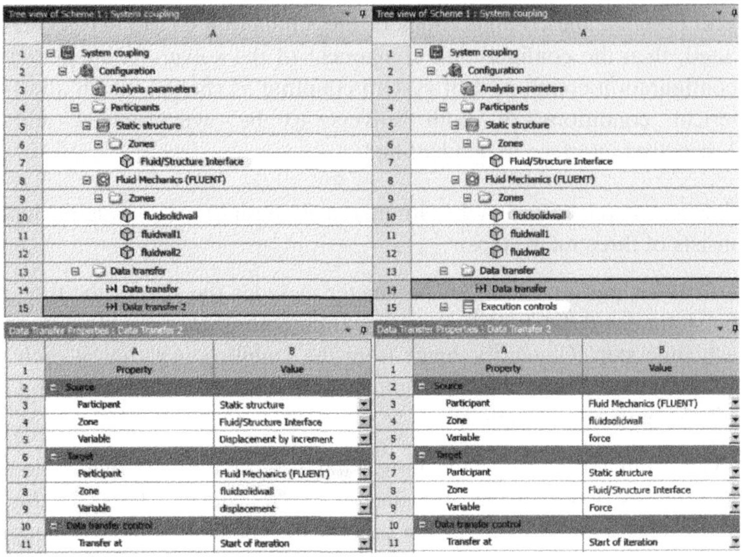

Figure 4.19. *Parameters of two-way coupling (left) and parameters of one-way coupling (right). For a color version of this figure, see www.iste.co.uk/elhami/interactions.zip*

4.8. Study of a 3D airplane wing

The numerical study proposed in this section presents an interaction problem involving a 3D airplane wing and an air flow. This application aims to illustrate the proposed methodology in the context of a deterministic analysis. The geometric and material properties of the coupled system are presented in Table 4.1:

Length	L = 10 m
Young's modulus	E = 7.1e10 Pa
Density of the structure	$\rho_s = 2,770$ kg.m^{-3}
Poisson coefficient	$\nu = 0.3$
Length	L = 12 m
Fluid density	$\rho_f = 1.225$ kg.m^{-3}
Viscocity	$\mu_f = 1.6e^{-5}$ kg/(m.s)

Table 4.1. *Geometric and material properties of the problem*

The total displacement of the wing subject to an airflow with initial speed X = 200 m/s calculated by Ansys/Fluent is presented in Table 4.2.

Ansys/Fluent	Total displacement (m)
One-way	0.759767
Two-way	0.77566

Table 4.2. *Total displacement*

Figures 4.21 and 4.22 show the pressure and the velocity of the fluid. The air passing under the wing is slowed, and the air that flows over the wing is accelerated. Bernouilli's theorem states that the pressure under the wing must therefore be higher than the pressure above the wing (the blue area in Figure 4.21). Thus, the lift force is the result of a pressure difference between the underside of the wing, called the *intrados*, and the top of the wing, the *extrados*. The magnitude of the lift force thus created depends on the pressure difference in the flow around the wing.

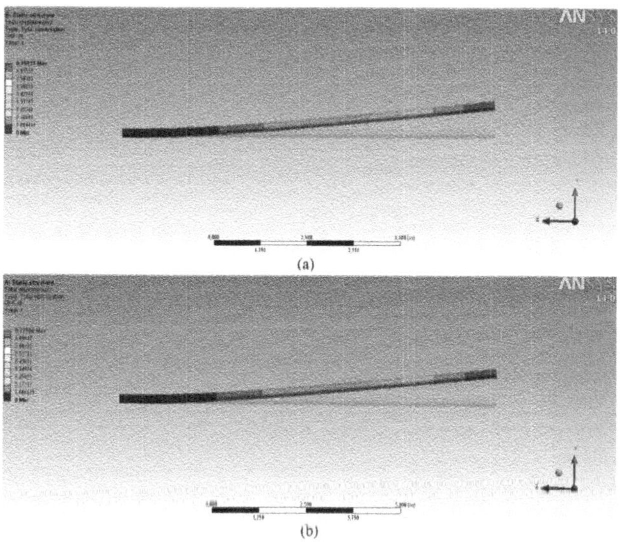

Figure 4.20. *Total displacement of the wing (a) one-way, b) two-way). For a color version of this figure, see www.iste.co.uk/elhami/Interactions.zip*

The total displacement of the wing subject to the air flow with initial speed VX = 200 m/s calculated by Ansys/Flotran is presented in Table 4.3.

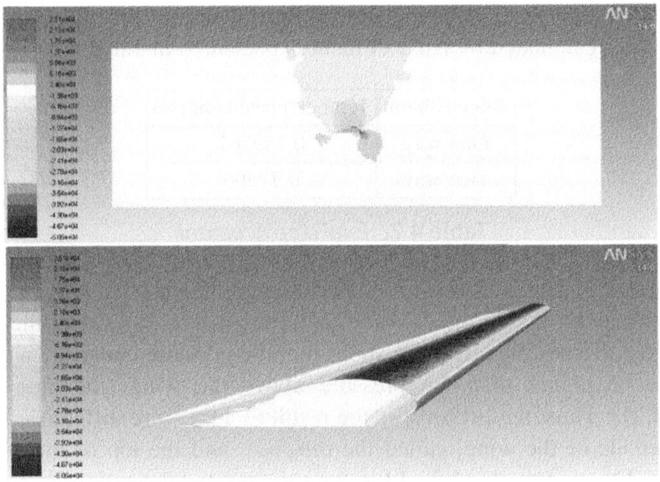

Figure 4.21. *Fluid pressure. For a color version of this figure, see www.iste.co.uk/elhami/interactions.zip*

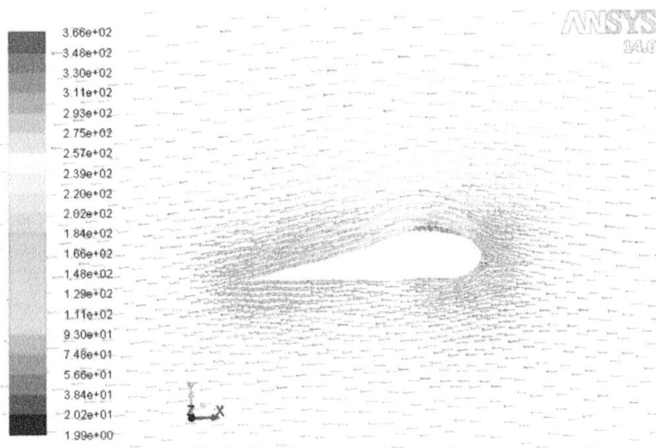

Figure 4.22. *Fluid velocity. For a color version of this figure, see www.iste.co.uk/elhami/interactions.zip*

	Ansys/Flotran
Total displacement (m)	0.662633

Table 4.3. *Total displacement*

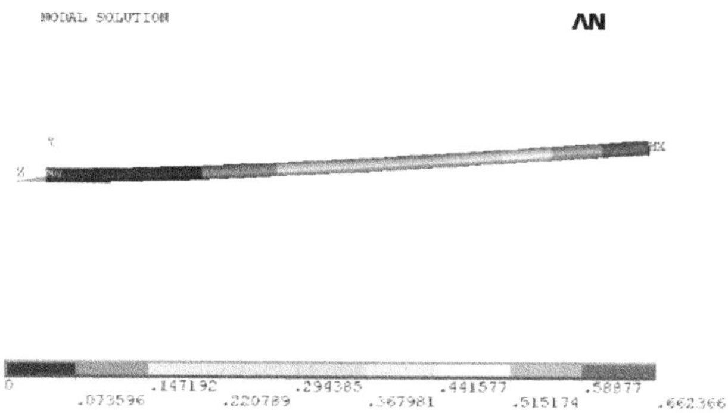

Figure 4.23. *Total displacement of the wing. For a color version of this figure, see www.iste.co.uk/elhami/interactions.zip*

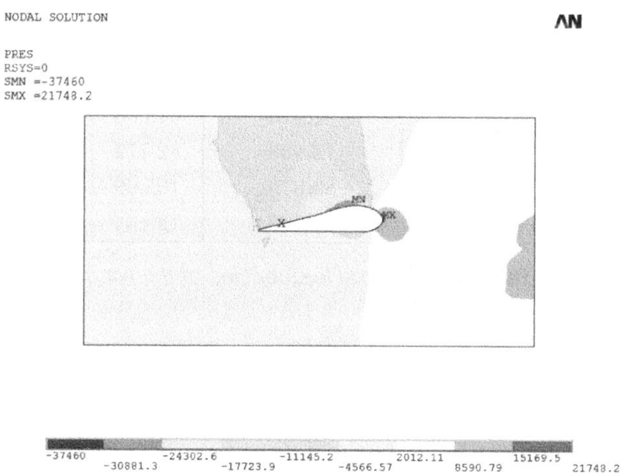

Figure 4.24. *Fluid pressure. For a color version of this figure, see www.iste.co.uk/elhami/interactions.zip*

4.8.1. *Modal analysis*

In general, modal analysis is used to determine the vibratory properties of a structure. These vibratory characteristics, such as the natural frequencies and modes, play an important role in designing structures that are subject to dynamic loads. It can be viewed as a starting point for transient dynamic analysis. The response of a structure to an excitation of its modes can also be evaluated. The fundamental equation used by undamped modal analysis is the classical eigenvalue problem given by:

$$[K]\phi_i = \lambda_i[M]\phi_i \qquad [4.116]$$

where $[K]$ is the stiffness matrix of the structure, ϕ_i is the eigenvector, λ_i is the eigenvalue and $[M]$ is the mass matrix of the structure. In prestressed modal analysis, the matrix $[K]$ incorporates the stress stiffness matrix. The results of modal analysis in both cases (Ansys/FLOTRAN and Ansys/Fluent [one-way and two-way]) are shown in Table 4.4. The results show the first six natural modes with their corresponding natural frequency values.

Frequencies	Without	With flow		
(Hz)	flow	Ansys/FLOTRAN	Ansys/Fluent	
			One-way	Two-way
F1	1.4830	12.630	14.640	14.853
F2	7.3146	19.182	22.016	22.290
F3	9.2564	51.426	61.008	62.014
F4	25.738	72.938	82.772	83.766
F5	29.314	89.873	105.69	107.36
F6	44.242	126.02	126.34	126.36

Table 4.4. *Natural frequencies of the wing*

4.9. Transient analysis

This application illustrates a two-way fluid–structure interaction problem in the context of the transient analysis of a 3D plate oscillating in a fluid-filled cavity. The two analysis systems need to be coupled throughout the whole solution to model the interaction between the structure and the fluid over time. The calculations are conducted in the regime of laminar flow.

The geometry is given by a closed 3D cavity of length 6 m and width 12 m and a thin plate of length 1 m fixed to the base of the cavity as shown in Table 4.5.

An initial pressure of 100 Pa is applied to one side of the thin plate for 0.5 s, inducing a deformation. When the pressure is released, the plate oscillates forwards and backwards to recover its (vertical) equilibrium position. The surrounding fluid damps the motion of the plate, which reduces the amplitude of the oscillations as a function of time. The Ansys/Fluent solver calculates the response of the fluid to the motion of the plate, and the Ansys/Mechanical solver calculates the deformation of the plate following the initial applied pressure resulting from the presence of the fluid. This coupling between the two solvers is necessary because the deformation of the structure affects the motion of the fluid, which in turn affects the deformation of the structure.

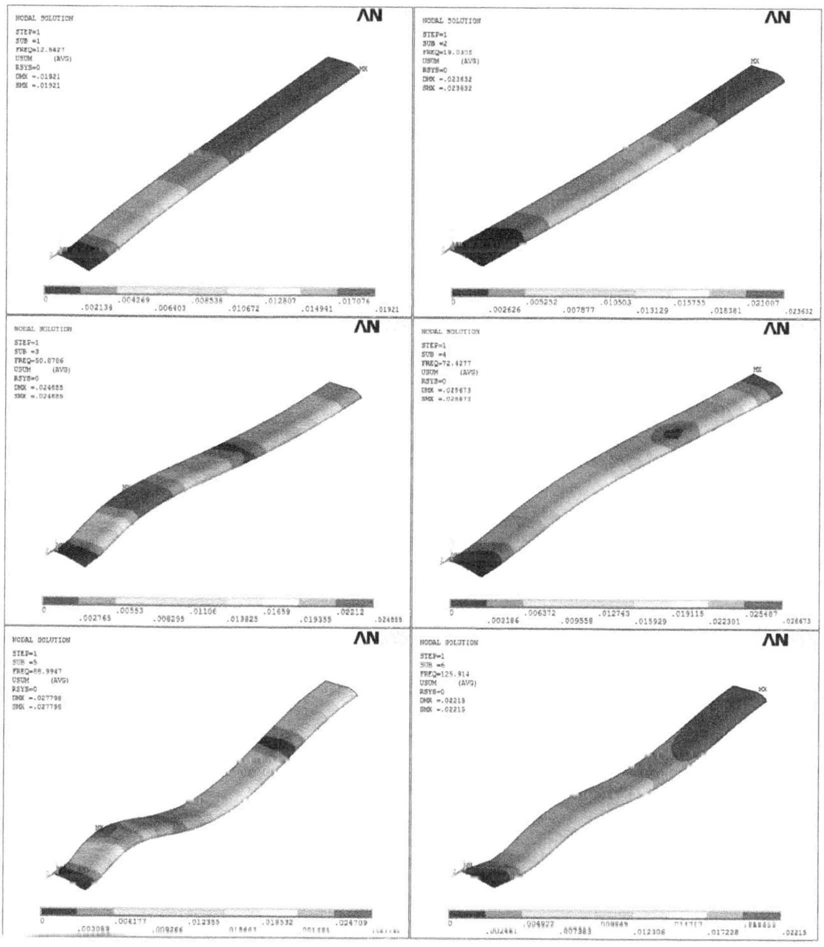

Figure 4.25. *Natural models of the deformed wing. For a color version of this figure, see www.iste.co.uk/elhami/interactions.zip*

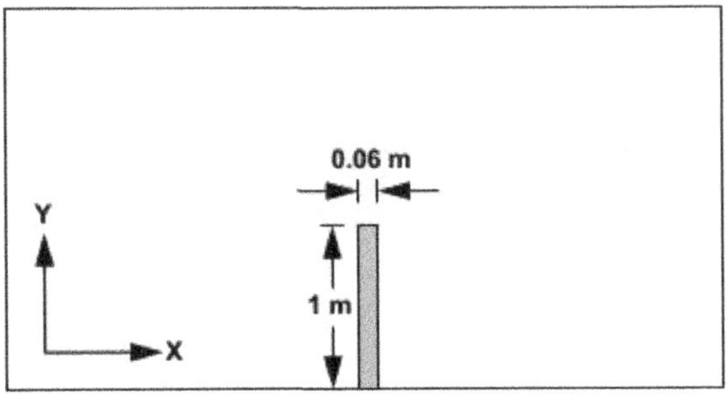

Figure 4.26. *Geometry of the model*

Length	L = 1 m
Thickness	H = 0.06 m
Depth	P = 0.4 m
Young's modulus	E = 2.5e6 Pa
Density of the structure	$\rho_s = 2,550$ kg.m^{-3}
Poisson coefficient	$\nu = 0.35$
Fluid density	$\rho_f = 1$ kg.m^{-3}
Dynamic viscosity	$\mu_f = 0.2$ kg/(m.s)

Table 4.5. *Geometric and material properties of the problem*

Two-way coupling has two separate data transfers:

– the data describing the force resulting from the motion of the air is received by the transient structural analysis system (Ansys/Mechanical), which solves the behavior of the structure over time;

– the data describing the displacement resulting from the motion of the plate is received by the fluid analysis system (Ansys/Fluent), which solves the behavior of the fluid over time.

The oscillation of the plate depends on time, so we must choose appropriate time values for the transient analysis:

– the total observed time is defined to be a period of 50 s, which is sufficient to observe multiple oscillation of the plate and to model the total damping as the plate returns to equilibrium;

– we define the time step to be 0.2 s, which is good enough to observe the oscillation to a reasonable degree. The choice of time step is extremely significant in transient analysis. If the time step is too large, the analysis will not see the behavior of the system, and if it is too small the computations will be too expensive.

The fluid and structure domains are meshed with hexahedral elements, with 40 elements for the structure mesh and 1,825 elements for the fluid mesh, as shown in Figures 4.27 and 4.28.

Figure 4.27. *Structure mesh. For a color version of this figure, see www.iste.co.uk/elhami/interactions.zip*

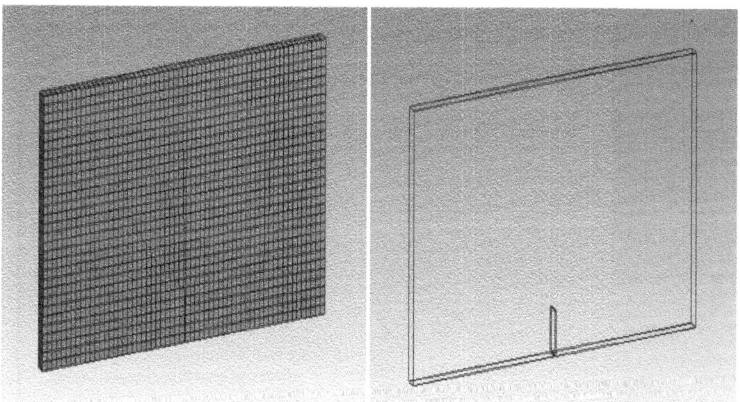

Figure 4.28. *Fluid mesh. For a color version of this figure, see www.iste.co.uk/elhami/interactions.zip*

To begin with, we are interested in the evolution of the free end of the plate. In Figure 4.29, we observe the position of this end over time. Figure 4.30 presents the evolution of the von Mises stress of the fixed end of the plate over time.

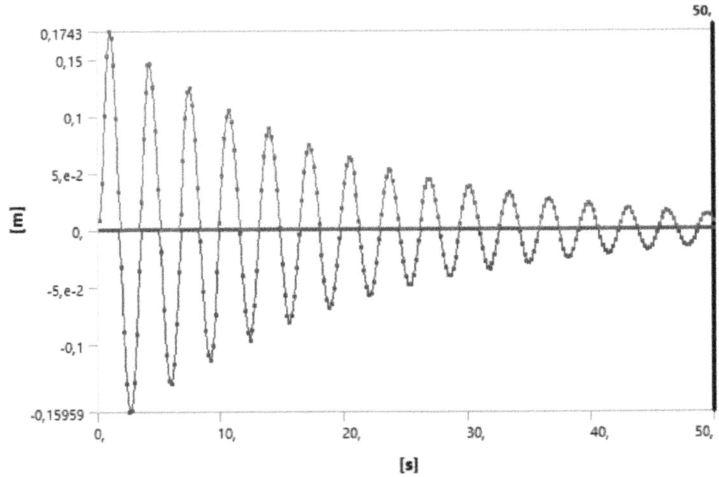

Figure 4.29. *Evolution over time of the displacement of the free end. For a color version of this figure, see www.iste.co.uk/elhami/interactions.zip*

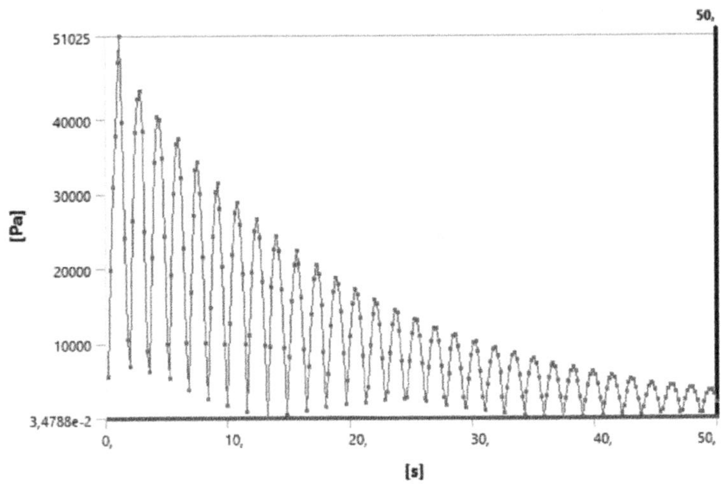

Figure 4.30. *Evolution over time of the von Mises stress. For a color version of this figure, see www.iste.co.uk/elhami/interactions.zip*

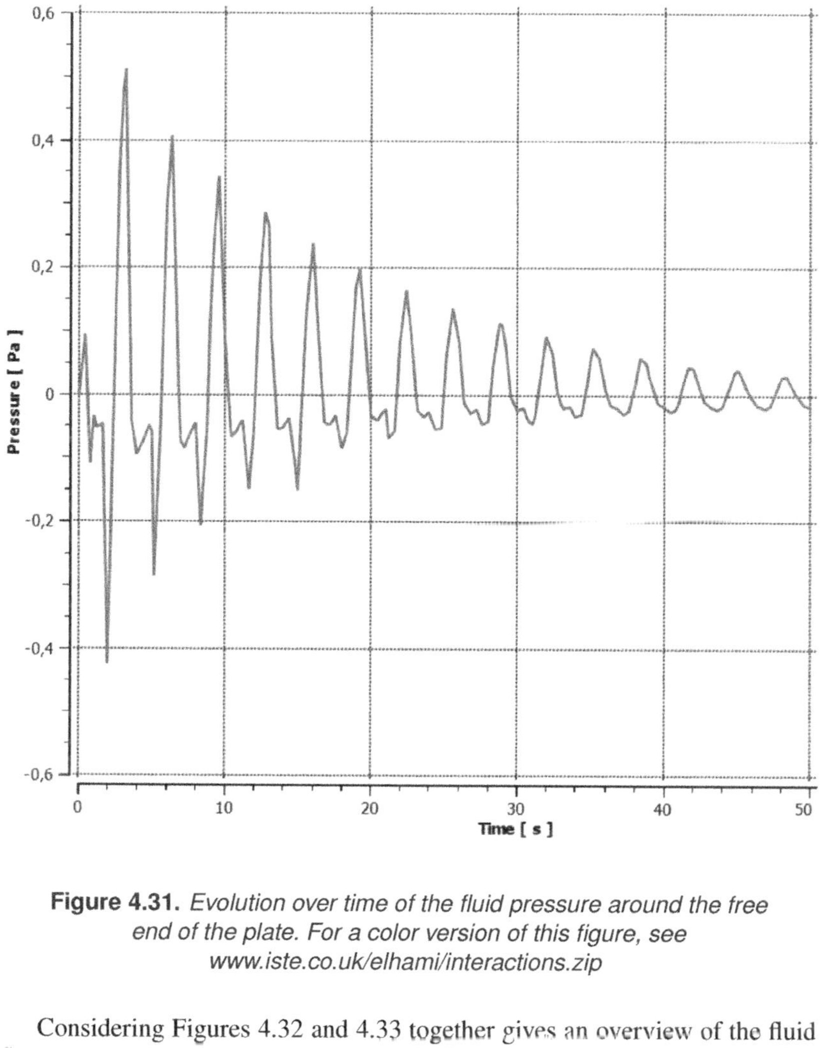

Figure 4.31. *Evolution over time of the fluid pressure around the free end of the plate. For a color version of this figure, see www.iste.co.uk/elhami/interactions.zip*

Considering Figures 4.32 and 4.33 together gives an overview of the fluid flow caused by the deformation of the beam over multiple characteristic periods of the IFS problem. These images show the pressure and the velocity field. Each of the vortices is visible, since the pressure there is minimal (in blue). We can see that several large vortices are released around the plate.

The elapsed physical time is 50 s with a time step of 0.2 s. The computation time was 50 min (on a PC Core-i7 2.2 GHz with 16 GB of RAM).

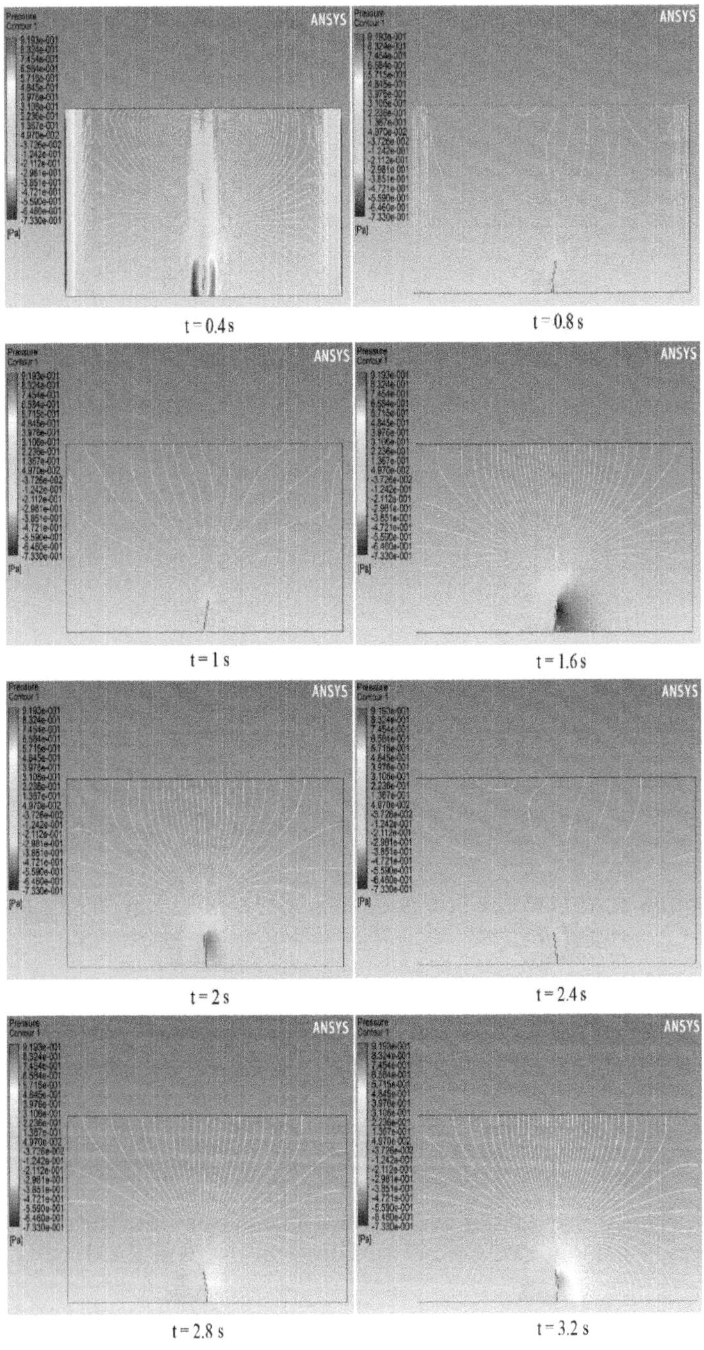

Figure 4.32. *Pressure field around the plate. For a color version of this figure, see www.iste.co.uk/elhami/interactions.zip*

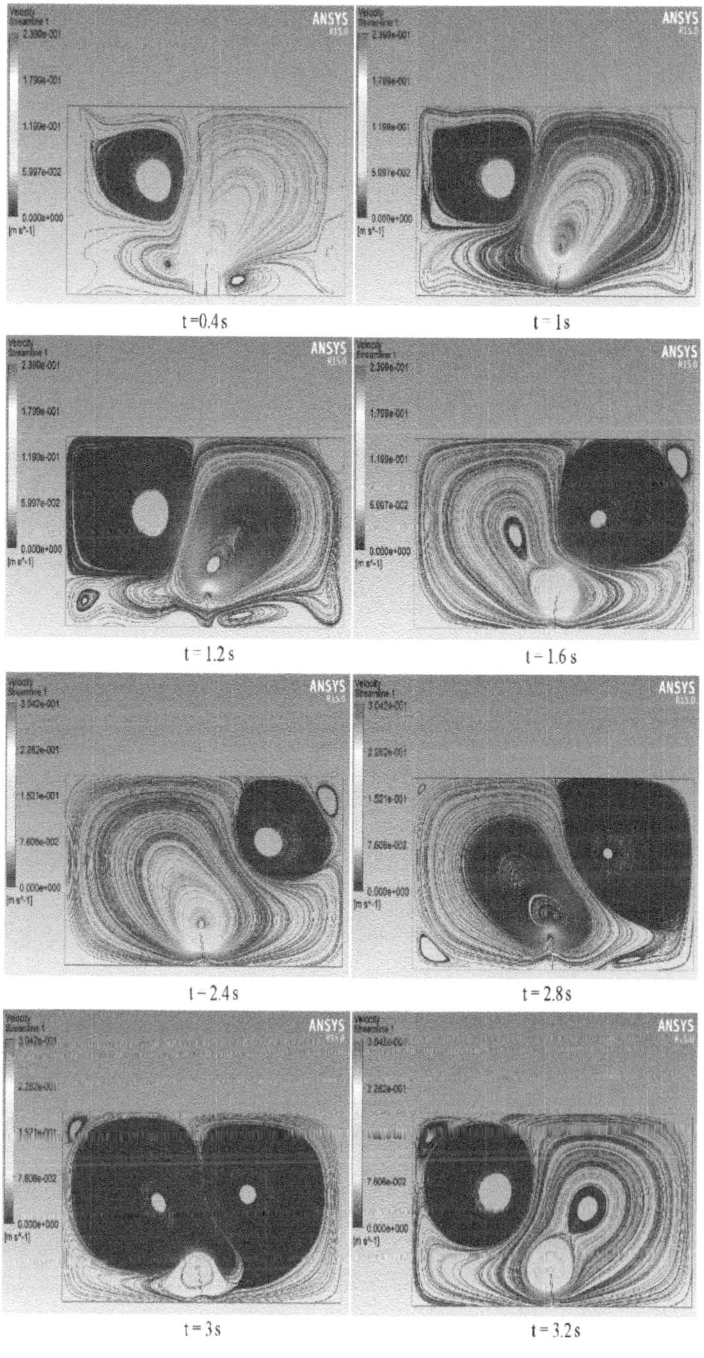

Figure 4.33. *Velocity field around the plate. For a color version of this figure, see www.iste.co.uk/elhami/interactions.zip*

5

Modal Reduction for FSI

5.1. Introduction

Numerical tools, long restricted by the lack of available computing power, are now widely used and are implemented with increasingly powerful models. For a given configuration, these numerical models can, for example, be used to determine the deformation of a solid and the physical properties of a flow. In industrial contexts, they allow the shape and the structure of materials to be optimized based on numerical results and make it possible to implement active monitoring.

However, the complexity of the phenomena that we are interested in studying translates into prohibitive computational costs, motivating us to search for reduced models with more realistic computation times. By a reduced model, we mean a low-dimensional description of a system obtained by analyzing classical numerical formulations. Finding this reduction incurs an initial cost, but this cost is largely offset if the reduced model is later found to be applicable for other configurations of parameters than those of the initial formulation. Furthermore, obtaining low-dimensional systems with fast computation times allows us to implement active monitoring, which is not possible with classical numerical models.

This chapter presents different methods for model reduction. We will begin by defining what we mean by the concept of model reduction. Indeed, any mathematical formulation with a simplified model for a given phenomenon or, analogously to data compression, any method that results in lower order models for problems in mechanics may be viewed as a model reduction.

These methods are based on the principle that each problem has an initial mathematical formulation given by finding $v(t)$ in an infinite-dimensional space $\mathcal{V}$:

$$A(v) = F \qquad [5.1]$$

where A is a differential operator that describes the studied phenomenon.

Numerically solving this problem involves projecting the system [5.1] onto the discrete basis of a space $\mathcal{V}^h \in \mathcal{V}$ with finite dimension $n = o\frac{1}{h}$, where h is the spatial discretization parameter: find $v^h \in \mathcal{V}^h$ such that

$$A^h(v^h) = F^h \qquad [5.2]$$

where A^h is the discretization of the operator A, and similarly for F^h.

This space $\mathcal{V}^h$ is a discretization of space $\mathcal{V}$, obtained for example by the finite element or finite volume methods. The numerical system that must be solved is now finite dimensional, but its dimension might still be extremely large, which would result in high computational costs.

The idea behind constructing a reduced model is to project either the continuous formulation (equation [5.1]) or the discrete formulation (equation [5.2]) onto a space $\tilde{\mathcal{V}}$ with dimension N much lower than $\mathcal{V}^h$ $(N \ll n)$, as is shown schematically in Figure 5.1, subject to the condition that $\tilde{\mathcal{V}}$ must retain certain properties, such as the energy of the solution, to within ε. Stochastic-based approaches to model reduction have been proposed in [BEN 09].

5.2. Dynamic substructuring methods

Complex structures often consist of multiple substructures. These substructures might be designed and developed by different teams working from different locations because of cost, labor, storage, etc. The substructuring method is therefore highly useful both in industrial contexts and in research. On the one hand, it significantly reduces the size of the problem and the computational cost. On the other hand, our understanding of the dynamic behavior of the global structure relies on our understanding of each substructure. Still, it is important to note that the substructuring method is very effective when the number of degree of freedom (DOF) of the coupling interface is much lower than that of each substructure.

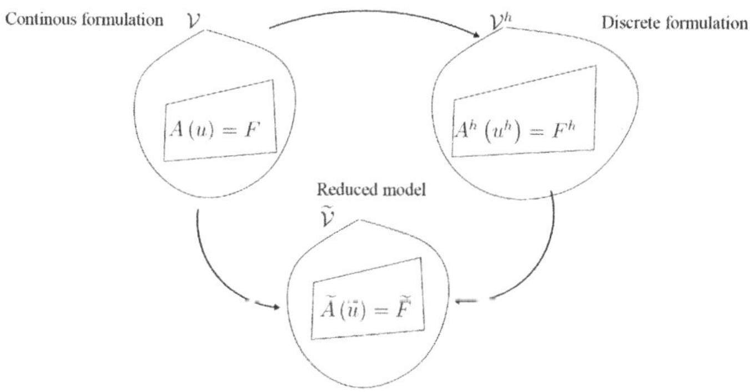

Figure 5.1. *Principle of model reduction*

Substructuring techniques are typically applied to structures with a linear behavior. When the behavior is nonlinear, they become more difficult to implement. Constructing a reduced model and finding the static deformations of each nonlinear substructure, however, is possible in some cases. In the next section, we will present a number of references from the literature on dynamic substructuring methods in the case of both linear and nonlinear problems.

5.2.1. *Linear problems*

The substructuring method was originally proposed by Hurty [HUR 65] in 1965. The initial version of the method considered substructuring with fixed coupling interfaces such that the dynamic behavior of each linear substructure is described by a set of modes:

1) Rigid body modes.

2) Static modes, defined as the deformations of the substructure arising from unit displacements of the points located on the coupling interface.

3) Normal modes of the substructure together with a fixed coupling interface.

The reaction forces at the coupling interface separate into two terms: statically determinate reactions and statically indeterminate reactions.

In 1968, Craig and Bampton [CRA 68] simplified the method for handling rigid body modes in Hurty's method by grouping together both of these reaction terms at the interface into a single term. Today, the Craig–Bampton substructuring method is one of the most widely used methods thanks to its simplicity, robustness and accuracy.

In parallel to this, other techniques have been proposed and developed based on Hurty's ideas. Depending on the boundary conditions applied to the coupling interfaces between the substructures, Tran [TRA 01] proposed classifying substructuring techniques into four groups:

– fixed interface (developed by [HUR 65] and [CRA 68]): this method is effective due to the simplicity of its implementation and the accuracy of its results. Explicit boundary conditions allow the elementary matrices to be directly assembled. However, they are difficult or sometimes even impossible to execute experimentally for verification by finite element modeling;

– free interface (developed by [MAC 71], [RUB 75] and [CRA 76]): this approach is also accurate. Unlike the above case, the free boundary conditions are implicit and are more manageable in the context of testing. However, the applicability of this method is limited due to its relatively complex formulation;

– hybrid interface (developed by [MAC 71] and [CRA 76]): since this method is a combination of the two previous methods, it inherits both of their disadvantages;

– loaded interface (developed by [BEN 71]): in this method, the coupling interfaces can be free or loaded, either partially or with respect to all DOFs. It is not often used, since it produces results with limited accuracy.

Depending on the domain of the problem, Girard and Roy [GIR 03] classify substructuring methods into three categories:

1) Matrix representation by directly assembling the mass, stiffness and damping matrices of each substructure:

This basic technique is suitable for small matrices.

2) Representation based on natural modes or modal synthesis:

This is well suited to the low-frequency range, where the principle of modal superposition is effective. This method reduces the model by using a basis consisting of truncated natural modes and static modes, which depend on the boundary conditions of the coupling interface, as described above. This is why Girard and Roy consider that the techniques associated with the four types of interface presented above belong to this category (modal synthesis).

3) Representation based on transfer functions or frequency synthesis:

This approach is simple to implement. It is very well suited to experiments, as transfer functions are very easily accessible features. These methods have a wider scope than the others, since industrial structures often have dynamic properties that depend on the frequency. For example, in soil–structure interaction problems, the impedance function of the soil depends on the oscillation frequency. However, the size of the problem can increase, since the transfer functions are calculated frequency by frequency, and some problems have a wide frequency band and the frequency discretization can be fine.

5.2.2. *Nonlinear problems*

Substructuring techniques for nonlinear problems can be classified into three categories:

1) Matrix representation:

As in the linear case, this basic technique is suitable for small systems. Because of the presence of nonlinear stiffness terms, it is clear that the computational cost of nonlinear problems is higher than that of linear problems.

2) Representation by natural modes:

Apiwattannalunggarn [API 03] applied the fixed interface substructuring method to nonlinear discrete systems with weak coupling between subsystems. The principle of Apiwattannalunggarn's approach is as follows:

– the equation describing the dynamics of each substructure is projected onto a linear Craig–Bampton basis, given by:

- natural modes, which are solutions of the vibration problem of the associated undamped linear substructure with fixed interface (fixed-interface problem);

- static modes calculated by imposing unit displacements to each DOF (static bearings problem)

– by using the invariant subspace idea by Shaw and Pierre [SHA 93], the elementary reduced model of each fixed-interface nonlinear substructure is determined. Note that it involves a nonlinear oscillator;

– using the displacement continuity condition and the stress equilibrium condition at the coupling interface, the elementary reduced nonlinear models are assembled to obtain a reduced global model.

For each nonlinear substructure, Iwatsubo *et al.* [IWA 03] used the multiple scales method to obtain a system of differential or linear algebraic equations, depending on the order of the perturbation. Finally, the authors apply the modal synthesis method to the full set of linear equations. They also propose to replace the method of multiple scales with the harmonic balance method, which allows linear equations to be obtained for the frequency components. These two methods are applied to a rotor system where the nonlinear internal force is of Duffing type (cubic in the displacement) and the external force is harmonic, with an excitation frequency close to the first critical frequency of the rotor. The authors conclude that the two proposed methods allowed good numerical results to be obtained even if only a few modes are considered, and that the method achieves a considerable reduction in the computation time.

Sundararajan *et al.* [SUN 98] used the fixed-interface substructuring method to obtain a reduced model of a rotor system consisting of linear subsystems with nonlinear interface elements. The vibration equation of the reduced global model was then solved by applying the shooting method combined with the continuation technique.

A similar approach [NGU 05] was adopted by replacing the nonlinear interface by nonlinear springs. The behavior of the subsystems remains linear. This approach was extended to the case of a beam whose behavior is assumed to be linear, coupled with a thin blade. The behavior of the blade is modeled by nonlinear springs and its mass is neglected, since it is much smaller than the mass of the beam. A reduced model is then obtained using the Craig–Bampton method. The experimental responses (from tests conducted at the University of Liège by the team led by Prof. J.C. Golinval) and the numerical responses obtained by directly integrating the equations of motion of the reduced model were compared using continuous wavelet analysis.

3) Representation based on transfer functions:

This type of method is well suited to nonlinear systems whose features depend on the frequency. Nonlinear substructuring methods based on frequency are explored in the thesis by Ferreira [FER 99]. Nonlinear differential equations are transformed into algebraic equations using two methods such as the multiharmonic balance method and the multiharmonic description functions method. The first is a generalized version of the harmonic balance method that defines higher order frequency response functions using Volterra series, whereas the second seeks to approximate (or linearize) the nonlinear forces using description functions. Finally, the subsystems were coupled using transfer functions.

5.3. Nonlinear substructuring method

Industrial problems in mechanics often involve multiple instances of nonlinearity. They are not only difficult to model, but are also difficult to solve, because:

– the principle of modal superposition no longer applies;

– the frequency varies as a function of the amplitude but also as a function of time.

The nonlinearities encountered in the structural vibrations arise from various different sources; they can either be the result of geometry, the constitutive law of the material, or contact or clearance phenomena between solids, etc. In general, it is very difficult to characterize which types of nonlinearity are present in the dynamic behavior of the structure based on experimental data alone. Classically, the characteristic nonlinearities of structural vibrations are often classified into three main categories:

– geometric nonlinearities: these are linked to the emergence of large amplitudes in the structural behavior. Since this is an example of large displacements, the stress–strain relationship ceases to be linear. Thin structures such as beams, plates and shells, nowadays heavily relied upon by the industry, are very susceptible to experiencing large displacements. Integrating geometric nonlinearities arising from large amplitudes into the behavioral models of thin structures is currently an extremely active area of international research;

– material nonlinearities: these originate from dislocations within the material itself. The relationship between the stress and the strain ceases to be linear. Examples of this behavior include elastoplasticity, nonlinear viscoelasticity, etc;

– contact nonlinearities: these arise at junctions between solids. They depend on contact, friction and clearance phenomena that unfold at the level of the links between solids.

Thin structures with geometric nonlinearities are now widely used in many different fields: civil engineering, aeronautics, etc. Attenuating and controlling the vibrations of these kinds of structures is crucial to avoid potentially dangerous phenomena during the normal usage conditions of these structures.

5.3.1. *Vibrational equations of a substructure*

We consider a structure composed of two substructures, (1) and (2). We will use the notation $(\cdot)^{(s)}$ $(s = 1, 2)$ to refer to each substructure. The vibration equation of each substructure is given by:

$$\begin{bmatrix} [M_{II}] & [M_{IB}] \\ [M_{BI}] & [M_{BB}] \end{bmatrix}^{(s)} \begin{bmatrix} \ddot{U}_I \\ \ddot{U}_B \end{bmatrix}^{(s)} + \begin{bmatrix} [K_{II}] & [K_{IB}] \\ [K_{BI}] & [K_{BB}] \end{bmatrix}^{(s)} \begin{bmatrix} U_I \\ U_B \end{bmatrix}^{(s)} + \begin{bmatrix} F_I(U) \\ F_B(U) \end{bmatrix}^{(s)} = \begin{bmatrix} 0 \\ G_B \end{bmatrix}^{(s)} \qquad [5.3]$$

where $[M]$, $[K]$ and F are the mass and stiffness matrices and the vector of nonlinear forces, respectively, which depend on the displacement U. The indices I and B, respectively, denote the internal DOF and the DOF of the coupling interface between the two substructures. G is the vector of coupling forces at the interface.

Note that the above equation has dimension $N = N_I + N_B$, where N_I and N_B, respectively, denote the number of internal DOF and the number of DOF of the coupling interface.

The assembly conditions at the coupling interface are given by:

– the continuity of the displacement:

$$U_B^{(1)} = U_B^{(2)} \qquad [5.4]$$

– the balance of coupling forces:

$$G_B^{(1)} + G_B^{(2)} = 0 \qquad [5.5]$$

In section 5.3.2, we will consider the two problems mentioned above: the fixed-interface problem and the static bearing problem.

5.3.2. *Fixed-interface problem*

The vibration modes of a substructure with a fixed interface, i.e. $U_B = 0$, are determined by solving the following equation:

$$[M_{II}]\ddot{U}_I + [K_{II}]U_I + F_I(U_I, U_B = 0) = 0 \qquad [5.6]$$

where the mass matrix $[M_{II}]$ and stiffness matrix $[K_{II}]$ are symmetric and positive definite.

This system is of size N_I. Equation [5.6] is projected onto the truncated linear modal basis consisting of the n first linear natural modes of the associated linear system: $[M_{II}]\ddot{\Phi}_I + [K_{II}]\Phi_I = 0$. The actual natural modes are easy to find: Φ_i (with $i = 1, \ldots, n$ and $n \leq N_I$).

The transformation

$$U_I(t) = [\Phi]q(t) = \sum_{i=1}^{n} \Phi_i q_i(t) \qquad [5.7]$$

gives us the modal representation of the vibration equation of the substructure:

$$\ddot{q}_i + \omega_i^2 q_i + f_i(q_j) = 0 \quad \text{with} \quad i, j = 1, \ldots, n \qquad [5.8]$$

where ω_i is the linear angular frequency and q_i are the modal coordinates.

In the classical linear case (without the $f_i(q_j)$ term), the Craig–Bampton method for the fixed-interface problem terminates after determining the reduced linear model. In the nonlinear case, we continue by applying the 1-mode and multiple-mode approaches proposed by Shaw and Pierre [SHA 93]. After choosing the set of "master" modes $k \subset S_M$, the other modal coordinates can be expressed in terms of the variables of these "master" modes (u_M, v_M):

$$q_i = X_i(\mathrm{u}_M, \mathrm{v}_M) \quad \dot{q}_i = Y_i(\mathrm{u}_M, \mathrm{v}_M) \quad \text{with } i = 1, \ldots, n \;\; i \notin S_M \qquad [5.9]$$

Choosing the number of "master" modes

Even though we are free to choose the elements from S_M, choosing insufficiently many elements or a poor configuration of elements can lead to an incorrect model. It is therefore preferable to begin by calculating the first frequencies up to a certain order for each of the associated linear substructures and for the associated global linear structure. After comparing the natural frequencies of the substructures with those of the global structure, we can deduce the frequencies (ω_C) that arise from assembling the substructures. The number of "master" modes (m) of the substructure (s) is then chosen so that the following condition is satisfied:

$$\omega_m^{(s)} \leq \max\left\{\omega_1^{(1)}\ \omega_1^{(2)}\ \cdots\ \omega_1^{(N_s)}\right\} \leq \omega_{m+1}^{(s)} \qquad [5.10]$$

where N_s is the number of substructures.

As we mentioned in the Introduction, it is preferable to choose: $S_M = [1, 2, \ldots, m]$ to obtain a good reduced model.

Note that to calculate the reduced model of the nonlinear substructures, we use Shaw and Pierre's 1-mode approach if $m = 1$, and their multiple-mode approach if $m > 1$.

5.3.3. *Static bearing problem*

In the static case, the equilibrium equation of each nonlinear substructure becomes:

$$\begin{bmatrix} [K_{II}] & [K_{IB}] \\ [K_{BI}] & [K_{BB}] \end{bmatrix} \begin{bmatrix} U_I \\ U_B \end{bmatrix} + \begin{bmatrix} F_I(U) \\ F_B(U) \end{bmatrix} = \begin{bmatrix} 0 \\ \tilde{G}_B \end{bmatrix} \qquad [5.11]$$

where $\tilde{G}_B$ is the static force at the coupling interface.

In the linear case, the static bearing problem leads to a linear relationship between the internal DOF and the DOF of the interface. This relationship is nonlinear if the behavior of the coupling interface is nonlinear. In general, the nonlinear static bearing problem does not have an analytical solution. Therefore, introducing nonlinear static modes makes the problem difficult to solve. Apiwattannalunggarn [API 03] assumes that the nonlinear part of the static modes is negligible compared to the linear part. The numerical examples given in the following sections show that the linear static modes are sufficient to obtain satisfactory results. We therefore have:

$$\begin{bmatrix} [K_{II}] & [K_{IB}] \\ [K_{BI}] & [K_{BB}] \end{bmatrix} \begin{bmatrix} U_I \\ U_B \end{bmatrix} \simeq \begin{bmatrix} 0 \\ \tilde{G}_B \end{bmatrix} \quad [5.12]$$

From the first line in the above equation, we obtain a relationship between the internal DOF and the DOF of the interface:

$$U_I = -[K_{II}]^{-1}[K_{IB}]U_B \quad [5.13]$$

5.3.4. *Representing the system with the linear Craig–Bampton basis*

The displacement of the internal DOF of a discrete subsystem may be written as:

$$U_I = U_{I,IF} + U_{I,RS} \quad [5.14]$$

where the displacements $U_{I,IF}$ and $U_{I,RS}$, whose expressions are given by equations [5.7] and [5.13], are, respectively, solutions of the fixed-interface problem and the static bearing problem. By introducing [5.7] and [5.13] into this equation, we therefore obtain:

$$U_I = [\Phi]q - [K_{II}]^{-1}[K_{IB}]U_B = [[\Phi] - [K_{II}]_{-1}[K_{IB}]] \begin{bmatrix} q \\ U_B \end{bmatrix} \quad [5.15]$$

The linear Craig–Bampton basis consists of:

1) natural modes, which are solutions of the fixed-interface problem;

2) static modes, which are solutions of the static bearing problem.

Projection onto the linear Craig–Bampton basis is achieved using the following transformation:

$$U = \begin{bmatrix} [\Phi] & -[K_{II}]^{-1}[K_{IB}] \\ [0] & [I] \end{bmatrix} \begin{bmatrix} q \\ U_B \end{bmatrix} \quad [5.16]$$

To simplify the notation, we set $U_C = U_B$. We use the index C to refer to the coupling interface. Using the transformation [5.16], the vibration equation [5.3] of the substructure (s) becomes:

$$\begin{bmatrix} [I] & [M_{SC}] \\ [M_{CS}] & [M_{CC}] \end{bmatrix}^{(s)} \begin{bmatrix} \ddot{q} \\ \ddot{U}_C \end{bmatrix}^{(s)} + \begin{bmatrix} [\Lambda_{SS}] & [0] \\ [0] & [K_{CC}] \end{bmatrix}^{(s)} \begin{bmatrix} q \\ U_C \end{bmatrix}^{(s)} + \begin{bmatrix} \tilde{F}_S(q, U^b) \\ \tilde{F}_C(q, U^b) \end{bmatrix}^{(s)} = \begin{bmatrix} 0 \\ G_C \end{bmatrix}^{(s)} \quad [5.17]$$

where the index S denotes the internal modal coordinates. $[\Lambda_{SS}] = \text{diag}\{\omega_1^2, \ldots, \omega_n^2\}$ is the matrix containing the squares of the linear angular frequencies of the substructure.

Now the equation obtained above, which has dimension $(n + N_C)$, is a reduced version of the original model (equation [5.3]) of dimension $(N_I + N_C)$, since $n < N_I$.

5.3.5. *Model reduction using the approach of Shaw and Pierre*

Omitting the (s) index of the substructures, we can expand system [5.17] as follows:

$$\ddot{q}_i + \sum_{s=1}^{N_c} [M_{SC}]_{is} \ddot{U}_s^C + \omega_i^2 q_i + \tilde{F}_i^S(q_j, q_k, U_l^C) = 0 \quad [5.18]$$

$$\ddot{q}_k + \sum_{s=1}^{N_c} [M_{SC}]_{ks} \ddot{U}_s^C + \omega_k^2 q_k + \tilde{F}_k^S(q_j, q_k, U_l^C) = 0 \quad [5.19]$$

$$\sum_{r=1}^{n} [M_{CS}]_{lr} \ddot{q}_r + \sum_{s=1}^{N_C} [M_{CC}]_{ls} \ddot{U}_s^C + \sum_{s=1}^{N_C} [K_{CC}]_{ls} U_s^C + \tilde{F}_l^C\left(q_j, q_k, U_l^C\right) = G_l^C \quad [5.20]$$

with $k \in S_M; \quad i, j = 1, ..., n; \quad i, j, r \notin S_M; \quad l, s = 1, ..., N_C$

Substituting [5.18] into [5.20], we therefore obtain:

$$\ddot{q}_k + \sum_{s=1}^{N_c} [M_{SC}]_{ks} \ddot{U}_s^C + \omega_k^2 q_k + \tilde{F}_k^S(q_j, q_k, U_l^C) = 0 \quad [5.21]$$

$$- \sum_{i=1, i \notin S_M}^{n} [M_{CS}]_{li} \left\{ \sum_{s=1}^{N_c} [M_{SC}]_{is} \ddot{U}_s^C + \omega_i^2 q_i + \tilde{F}_i^S(q_j, q_k, U_l^C) \right\}$$

$$+ \sum_{k=1, k \in S_M}^{n} [M_{CS}]_{lk} \ddot{q}_k + \sum_{s=1}^{N_c} [M_{CC}]_{ls} \ddot{U}_s^C + \sum_{s=1}^{N_c} [K_{CC}]_{ls} U_s^C$$

$$+ \tilde{F}_l^C(q_j, q_k, U_l^C) = G_l^b \quad [5.22]$$

with $k \in S_M$; $i, j = 1, \ldots, n$; $i, j, r \notin S_M$ and $l, s = 1, \ldots, N_C$.

Equation [5.22] can be rearranged as follows:

$$\sum_{k=1, k \in S_M}^{n} [M_{CS}]_{lk} \ddot{q}_k + \sum_{s=1}^{N_c} [M_{CC,\text{new}}]_{ls} \ddot{U}_s^C + \sum_{s=1}^{N_c} [K_{CC}]_{ls} U_s^C$$

$$+ \tilde{F}_l^{C,\text{new}}(q_j, q_k, U_l^C) = G_l^C \quad [5.23]$$

with:

$$[M_{CC,\text{new}}]_{ls} = [M_{CC}]_{ls} - \sum_{i=1, i \notin S_M}^{n} [M_{CS}]_{li} [M_S C]_{is} \quad [5.24]$$

$$\tilde{F}_l^{C,\text{new}} = \tilde{F}_l^C - \sum_{i=1, i \notin S_M}^{N_c} [M_{CS}]_{li} \left\{ \omega_i^2 q_i + \tilde{F}_i^S(q_j, q_k, U_l^C) \right\} \quad [5.25]$$

$k \in S_M$; $i, j = 1, \ldots, n$; $i, j, r \notin S_M$ and $l, s = 1, \ldots, N_C$.

In the fixed-interface problem, the multiple-mode approach by Shaw and Pierre was used to calculate the expressions of the q_i as a function of the m-couples of variables (u_M, v_M) of the "master" modes S_M, i.e.

$$q_i = X_i(\mathrm{u}_M, \mathrm{v}_M) \quad [5.26]$$

$$\dot{q}_i = Y_i(\mathrm{u}_M, \mathrm{v}_M) \quad [5.27]$$

with $i = 1, \ldots, n$; $i \notin S_M$ (where $i > m$).

Recall that $u_M = q_M$, $v_M = \dot{q}_M$, where $q_M = [q_1, q_2, \ldots, q_m]^T$ and m is the number of "master" modes.

Regrouping together [5.21] and [5.22], and taking into account [5.26] and [5.27], we obtain a single equation in matrix form with variables given by $(q_M, \dot{q}_M, U_C)$:

$$\begin{bmatrix} [I_{MM}] & [M_{MC}] \\ [M_{CM}] & [M_{CC,\text{new}}] \end{bmatrix} \begin{bmatrix} \ddot{q}_M \\ \ddot{U}_C \end{bmatrix} + \begin{bmatrix} [\Lambda_{MM}] & [0] \\ [0] & [K_{CC}] \end{bmatrix} \begin{bmatrix} q_M \\ U_C \end{bmatrix} + \begin{bmatrix} \tilde{F}_M(q_M, \dot{q}_M, U_C) \\ \tilde{F}_{C,\text{new}}(q_M, \dot{q}_M, U_C) \end{bmatrix} = \begin{bmatrix} 0 \\ G_C \end{bmatrix} \quad [5.28]$$

where $[I_M M]$ is the identity matrix of size $m \times m$. $[\Lambda_{MM}] = \text{diag}\{\omega_1^2, \ldots, \omega_m^2\}$ contains the first m angular frequencies (squared) of the substructure. $\tilde{F}_M = \left[\tilde{F}_1, \ldots, \tilde{F}_m\right]^T$ is the vector of nonlinear internal forces.

In summary, the change in the order of the reduced model of the substructure (s) can be shown as follows:

– start (without reduction): $N = N_I + N_C$, where N_I and N_C, respectively, represent the number of internal DOF and the number of DOF at the coupling interface;

– after projecting on the linear Craig–Bampton basis: $N_1 = n + N_C$ with $n < N_I$, where n is the chosen number of linear natural modes;

– finally, after applying Shaw and Pierre's multiple-mode (or 1-mode) approach: $N_2 = m + N_C$, with $m < n$, where m is the number of "master" modes.

5.3.6. *Assembling the substructures*

Switching notations, the assembling conditions [5.4] and [5.5] become:

$$\begin{cases} U_C^{(1)} = U_C^{(2)} = U_C \\ G_C^{(1)} = G_C^{(2)} = 0 \end{cases} \quad [5.29]$$

Equation [5.28] is rewritten by introducing the notation (s):

$$\begin{bmatrix} [I_{MM}] & [M_{MC}] \\ [M_{CM}] & [M_{CC,\text{new}}] \end{bmatrix}^{(s)} \begin{bmatrix} \ddot{q}_M \\ \ddot{U}_C \end{bmatrix}^{(s)} + \begin{bmatrix} [\Lambda_{MM}] & [0] \\ [0] & [K_{CC}] \end{bmatrix}^{(s)} \begin{bmatrix} q_M \\ U_C \end{bmatrix}^{(s)} + \begin{bmatrix} \tilde{F}_M(q_M, \dot{q}_M, U_C) \\ \tilde{F}_{C,\text{new}}(q_M, \dot{q}_M, U_C) \end{bmatrix}^{(s)} = \begin{bmatrix} 0 \\ G_C \end{bmatrix}^{(s)} \quad [5.30]$$

Using the assembly conditions stated above, we obtain:

$$\begin{bmatrix} [I_{MM}]^{(1)} & 0 & [M_{MC}]^{(1)} \\ [0] & [I_{MM}]^{(2)} & [M_{MC}]^{(2)} \\ [M_{CM}]^{(1)} & [M_{CM}]^{(2)} & [M_{CC,\text{new}}] \end{bmatrix} \begin{bmatrix} \ddot{q}_M^{(1)} \\ \ddot{q}_M^{(2)} \\ \ddot{U}_C \end{bmatrix} + \begin{bmatrix} [\Lambda_{MM}]^{(1)} & [0] & [0] \\ [0] & [\Lambda_{MM}]^{(2)} & [0] \\ [0] & [0] & [K_{CC}] \end{bmatrix} \begin{bmatrix} q_M^{(1)} \\ q_M^{(2)} \\ U_C \end{bmatrix} + \begin{bmatrix} \tilde{F}_M^{(1)}(q_M^{(1)}, \dot{q}_M^{(1)}, U_C) \\ \tilde{F}_M^{(2)}(q_M^{(2)}, \dot{q}_M^{(2)}, U_C) \\ \tilde{F}_{C,\text{new}}(q_M, \dot{q}_M, U_C) \end{bmatrix} = \begin{bmatrix} 0 \\ 0 \\ 0 \end{bmatrix}^{(s)} \quad [5.31]$$

with

$$\begin{aligned} q_M &= \begin{bmatrix} q_M^{(1)} \\ q_M^{(2)} \end{bmatrix} \\ &= [M_{CC,\text{new}}]^{(1)} + [M_{CC,\text{new}}]^{(2)} \\ &= [K_{CC}]^{(1)} + [K_{CC}]^{(2)} \\ \tilde{F}_{C,\text{new}}(q_M, \dot{q}_M, U_C) &= \tilde{F}_{C,\text{new}}^{(1)}\left(q_M^{(1)}, \dot{q}_M^{(1)}, U_C\right) + \tilde{F}_{C,\text{new}}^{(2)}\left(q_M^{(2)}, \dot{q}_M^{(2)}, U_C\right) \end{aligned}$$

Equation [5.31] characterizes the reduced model of the global structure. Note that the order of this model is smaller than that of the full model (without reduction).

5.4. Proper orthogonal decomposition for flows

Reduction methods are frequently used in fluid mechanics and structural mechanics. In fluid mechanics, proper orthogonal decomposition (POD) has been successfully applied to problems since the 1990s and has replaced other methods. Formulated as an eigenvector problem for a spatial correlation matrix constructed from snapshots of the flow, it allows a basis for a velocity field to be constructed that is optimal with respect to energy. In structural mechanics, the POD basis is comparable to a modal basis for vibrating linear structures. However, it has only been very infrequently applied to fluid–structure interaction problems, since the spatial nature of the basis creates difficulties when considering mobile domains.

POD was introduced in fluid mechanics by Lumley in 1967 with the goal of identifying coherent structures within turbulent flows [LUM 67]. In this context, POD is applied by a method that we describe here as the classical method, which finds the eigenvectors of a spatial correlation operator of the velocity field. However, it was not until the POD approach was complemented by the Snapshot method [SIR 87], which reduces the size of the problem to be solved, that POD began to be further explored in fluid mechanics. Since the 1990s, POD has been used in a wide range of applications for various different problems in fluid mechanics.

Aubry *et al.* [AUB 88] were the first to apply a reduced POD model to the study of the turbulent boundary layer. Since then, various different configurations have been studied, such as the evolution of the turbulent boundary layer along a flat plate [REM 94, REM 96], channeled flows [OMU 99] and dragged cavities [CAZ 98]. A very important adjustment to this method was also made in aerodynamics in order to study shock waves [LEG 01, LIA 03].

In general, there are two possible interpretations of POD. The first interpretation views POD as a Karhunen–Loeve decomposition (KLD) [NEW 96], or more specifically as a set of three methods: KLD, which was developed for stochastic processes, principal component analysis from data analysis and singular value decomposition, which is used in the case of non-square matrices. Liang *et al.* [LIA 02] and Wu *et al.* [WU 03] compared these three methods and showed that they lead to the same formulation, i.e. an eigenvector problem for a spatial correlation operator.

5.4.1. *Properties of POD modes*

The basis obtained by this method has the following properties:

– the modes Φ satisfy homogeneous boundary conditions. In the case of an incompressible fluid, the velocity field has zero divergence, and so does the POD basis:

$$\text{div}\Phi_i = 0 \qquad [5.32]$$

– the time coefficients $a_i(t)$ can be obtained by projecting the velocity field v onto the basis (Φ_i):

$$a_i(t) = (v(t), \Phi_i) \qquad [5.33]$$

– the coefficients are independent and their eigenvalues are equal to their average over time:

$$\langle a_i(t)a_j(t)\rangle = \delta_{ij}\lambda_i \qquad \text{without summing over repeated indices} \quad [5.34]$$

– each eigenvalue λ_i represents the proportion of energy captured by the mode Φ_i. This means that if we truncate the basis to N modes, $N < n_R$, then the decomposition

$$v(x,t) - \sum_{i=1}^{N} a_i(t)\Phi_i(x) \qquad [5.35]$$

is optimal in terms of energy. It is not possible to obtain a decomposition into N nodes with higher energy than the decomposition [5.35].

– the error induced by the basis is equal to the sum of the eigenvalues of the neglected modes:

$$\langle \|v(t) - \sum_{k=1}^{N} a_k(t)\Phi_k\|^2\rangle = \sum_{k=N+1}^{n_R} \lambda_k \qquad [5.36]$$

5.4.2. *Snapshot POD*

For problems with large dimensions, finding the eigenvectors of the operator R can prove expensive. Indeed, we must solve a problem of dimension $4n_{x_1}n_{x_2}$ in 2D or $9n_{x_1}n_{x_2}n_{x_3}$ in 3D, where n_{x_1}, n_{x_2}, n_{x_3} are the number of mesh nodes along the axes x_1, x_2 and x_3. The computational power available from a single computer will quickly be overwhelmed. To remedy this, Sirovitch introduced the Snapshots method [SIR 87].

This method assumes that if the flow is accurately described by M snapshots, $M \ll nn_c$, where n is the number of nodes and nc is the number of components, then the problem to be solved does not have dimension $nn_c \times nn_c$ but rather $M \times M$.

The idea is to not attempt to find the basis (Φ_i), $i = 1, \ldots, n_R$, directly, but instead to first look for coefficients A_k, $k = 1, \ldots, M$ such that:

$$\Phi_k(x) = \sum_{i=1}^{M} A_i^k v(x, t_i) \qquad [5.37]$$

The problem is thus reduced to the following:

$$\sum_{i=1}^{M} = \frac{1}{M}(v(t_i), v(t_j))A_j^k = \lambda_k A_k^i \quad \text{for } i = 1...M \qquad [5.38]$$

where λ_k is the eigenvalue associated with the POD mode Φ_k.

The POD basis is then obtained by calculating [5.37] and the time coefficients a_i are found using equation [5.33]. There are no fixed rules for the number of required snapshots or the intervals between consecutive snapshots.

The choice between the snapshot method and the classical approach is made based on the nature of the data being processed. In the case of numerical simulations with large spatial meshes and small time samples, the snapshot method is recommended.

5.4.3. *Finding low-order expressions for dynamic systems*

POD allows us to find a basis that is optimal with respect to energy, i.e. such that the first vectors contain the largest proportion of the energy of the field (in a statistical sense). It is reasonable to expect that a Galerkin projection of the Navier–Stokes equations onto the first N energy modes will allow us to obtain a low-order dynamic system that correctly reproduces the studied phenomenon. Since the basis is spatial and therefore fixed in time, performing a Galerkin projection of the Navier–Stokes equations yields an ordinary system of equations in time that is much faster to solve.

5.4.3.1. *Dynamic system based on the instantaneous velocity field*

Consider the fluid velocity field v projected onto the POD basis truncated to N nodes:

$$v_N(x,t) = \sum_{k=1}^{N} a_k(t)\Phi_k(x) \qquad [5.39]$$

In the remainder of this section, we will use the same notation for v and v_N. Recall the Navier–Stokes equations for an incompressible fluid in dimensionless form:

$$\begin{cases} \nabla \cdot v & = 0 \\ \frac{\partial v}{\partial t} + v \cdot \nabla v & = -\nabla p + \frac{1}{Re}\Delta v \end{cases} \qquad [5.40]$$

where Re is the dimensionless Reynolds number, ρ is the density of the fluid and μ is its dynamic viscosity.

For incompressible flows, the POD basis has zero divergence. Therefore, we do not need to take the incompressibility equation into account.

Consider the weak formulation of equation [5.40] in terms of the POD basis:

$$\begin{cases} \text{Find } v \in \tilde{V} \text{ such that } \nabla \cdot v = 0 \\ \forall \Phi_i, i = 1, \dots, N, \ \nabla \cdot \Phi = 0 \\ \displaystyle\int_\Omega \frac{\partial v}{\partial t} \cdot \Phi_i dx + \int_\Omega (v \cdot \nabla v) \cdot \Phi_i dx \\ \quad \displaystyle = -\int_\Omega \nabla p \cdot \Phi_i dx + \frac{1}{Re}\int_\Omega \Delta v \cdot \Phi_i dx \end{cases} \qquad [5.41]$$

From the decomposition [5.39] of v over the basis truncated to N modes, and from the orthogonality of the POD modes, we obtain the following system:

$$\frac{da_i}{dt} = \sum_{k=1}^{N}\sum_{l=1}^{N} a_k a_l C_{kli} + \sum_{k=1}^{N} a_k B_{ki} + D_i \qquad \text{with } i = 1 \cdots N \qquad [5.42]$$

where

$$\begin{aligned} C_{kli} &= -(\Phi_k \cdot \nabla\Phi_l, \Phi_i) = -\int_\Omega (\Phi_k \cdot \nabla\Phi_l) \cdot \Phi_i dx \\ B_{ki} &= \tfrac{1}{Re}(\Delta\Phi_k, \Phi_i) = \tfrac{1}{Re}\int_\Omega \Delta\Phi_k \cdot \Phi_i dx \\ D_i &= -\int_{\partial\Omega} p\Phi_i \cdot \mathbf{n} dx \end{aligned}$$

and where $\mathbf{n}$ is the outward normal of the fluid domain on the boundary $\partial\Omega$.

To solve this system, it is sufficient to precalculate the coefficients C_{kli}, B_{ki} and D_i. Since the POD basis was for the velocity, D_i is a problematic term that remains to be calculated.

5.4.3.2. *Dynamic system based on the fluctuating velocity field*

In practice, the first POD function captures the essential part of the signal, so the velocity field is often decomposed into a steady part and a fluctuating part. POD is then applied to the fluctuating field, allowing the following dynamic system to be constructed:

$$v = \langle v \rangle + v' \qquad p = \langle p_i \rangle + p' \qquad [5.43]$$

where $\langle \cdot \rangle$ is the time averaging operator detailed above. For convenience when writing the equations, we will henceforth replace $\langle \cdot \rangle$ by $\bar{\cdot}$.

This equation is substituted into equation [5.40], and its average is subtracted from the result. We thus obtain the Navier–Stokes equation for fluctuating quantities:

$$\frac{\partial v'}{\partial t} + v' \cdot \nabla v' + \bar{v} \cdot \nabla v' + v' \cdot \nabla \bar{v} - \overline{v' \cdot \nabla v'} = -\nabla p' + \frac{1}{Re}\Delta v' \qquad [5.44]$$

Proceeding in the same way as for the instantaneous field, we decompose v' over the POD basis Φ' truncated to N modes:

$$v'(x,t) = \sum_{k=1}^{N} a'_k(t)\Phi'_k(x) \qquad [5.45]$$

We thus obtain the dynamic system for the fluctuating field:

$$\frac{da'_i}{dt} = \sum_{k=1}^{N}\sum_{l=1}^{N}(a'_k a'_l - \overline{a'_k a'_l})C_{kli} + \sum_{k=1}^{N} a'_k B_{ki} + D_i \qquad [5.46]$$

where

$$\begin{aligned} C_{kli} &= -\int_\Omega(\Phi'_k \cdot \nabla\Phi'_l)\cdot\Phi'_i dx \\ B_{ki} &= \tfrac{1}{Re}\int_\Omega \Delta\Phi'_k \cdot \Phi'_i dx - \int_\Omega(\nabla\bar{v}\cdot\Phi'_k)\cdot\Phi'_i dx - \int_\Omega(\nabla\Omega'_k\cdot\bar{v})\cdot\Phi'_i dx \\ D_i &= -\int_{\partial\Omega} p'\Phi'_i \cdot \mathbf{n} dx \end{aligned}$$

There are several ways of calculating the term $\overline{a'_k a'_l}$. The first method is to estimate this term during computations based on the previous time steps. The second method is to assume that the term is constant with respect to time and use the following equation:

$$\overline{a'_k a'_l} = \lambda_k \delta_{kl} \qquad [5.47]$$

A third possibility was suggested by Allery [ALL 02], who obtained the following dynamic system by directly introducing [5.43] into [5.40]:

$$\frac{da'_i}{dt} = \sum_{k=1}^{N}\sum_{l=1}^{N} a'_k a'_l C_{kli} + \sum_{k=1}^{N} a'_k B_{ki} + D_i + H_i \qquad [5.48]$$

where

$$C_{kli} = -\int_\Omega (\Phi'_k \cdot \nabla\Phi'_l)\cdot\Phi'_i dx$$

$$B_{ki} = \frac{1}{Re}\int_\Omega \Delta\Phi'_k \cdot \Phi'_i dx - \int_\Omega(\nabla\bar{v}\cdot\Phi'_k)\cdot\Phi'_i dx - \int_\Omega(\nabla\Phi'_k\cdot\bar{v})\cdot\Phi'_i dx$$

$$D_i = -\int_{\partial\Omega} p'\Phi'i \cdot \mathbf{n}dx$$

$$H_i = -\int_{\Omega} \left(\nabla\bar{p} - \frac{1}{Re}\Delta\bar{v} + \bar{v}\cdot\nabla\bar{v}\right)\cdot\Phi_i' dx$$

5.4.3.3. *Handling the pressure term*

The reduced dynamic systems obtained thus far include a pressure term that cannot be projected onto the POD modes. For flows with homogeneous boundary conditions, since the POD modes cancel wherever the velocity field is zero, the coefficients D_n are not taken into account. In the other cases, we must either model the term [AZR 99], or attempt to eliminate it. Rempfer [REM 96] proposes to either use a Poisson equation to relate the pressure to the velocity, or to apply POD to the vorticity Navier–Stokes equations to eliminate the pressure. Allery [ALL 02] instead proposes to use a stress-based formulation. This final formulation is the one that we will explore below.

The method transforms the boundary conditions for the velocity into boundary conditions for the stress of type $\sigma \cdot \mathbf{n} = F$. The boundary of the fluid domain Γ_f is decomposed into a subboundary Γ_v where the velocity is zero and a subboundary Γ_σ on which a boundary condition for the stress is imposed:

$$\Gamma_f = \Gamma_v \cup \Gamma_\sigma \qquad [5.49]$$

Then, by considering v^{cl}, the velocity on Γ_σ, and $v|_{\Gamma_\sigma}$, the value of the velocity calculated on this same boundary, we introduce the following relation:

$$\sigma \cdot \mathbf{n}|_{\Gamma_\sigma} = F = G\left(v|_{\Gamma_\sigma} - v^{cl}\right) \qquad [5.50]$$

where G is a constant. We therefore have:

$$v|_{\Gamma_\sigma} - v^{cl} = \frac{F}{G} \qquad [5.51]$$

$v|_{\Gamma_\sigma}$ tends to v^{cl} as G becomes large relative to $\|F\|$. This technique therefore penalizes the Dirichlet boundary conditions. It was introduced by Batoz [BAT 90] for finite elements.

The dynamic systems are formulated by keeping the right-hand side $\nabla \cdot \sigma$ and replacing $\sigma \cdot \mathbf{n}$ in the integral $\int_{\Gamma_f} (\sigma \cdot \mathbf{n}) \cdot \Phi dx$ by $G\left(v|_{\Gamma_\sigma} - v^{cl}\right)$, which gives the following for the dynamic system for the instantaneous velocity field:

$$\frac{da_i}{dt} = \sum_{k=1}^{N}\sum_{l=1}^{N} a_k a_l C_{kli} + \sum_{k=1}^{N} a_k(B_{ki} + E_{ki}) + D_i \text{ with } i = 1 \cdots N \quad [5.52]$$

where

$$\begin{aligned} C_{kli} &= -(\Phi_k \cdot \nabla\Phi_l, \Phi_i) \\ B_{ki} &= -\tfrac{1}{\mathrm{Re}}(\mathbf{Tr}(\mathbf{D}(\Phi_k)), \nabla\Phi_i) \\ D_i &= -G \int_{\Gamma_\sigma} v^{cl}\Phi_i \cdot \mathbf{n} dx \\ E_{ki} &= G \int_{\Gamma_\sigma} \Phi_k \cdot \Phi_i dx \end{aligned}$$

For the fluctuating velocity field, we have:

$$\frac{da_i'}{dt} = \sum_{k=1}^{N}\sum_{l=1}^{N} a_k' a_l' C_{kli} + \sum_{k=1}^{N} a_k'(B_{ki} + E_{ki}) + D_i + K_i \quad [5.53]$$

where

$$\begin{aligned} C_{kli} &= -(\Phi_k' \cdot \nabla\Phi_l', \Phi_i') \\ B_{ki} &= -\tfrac{1}{\mathrm{Re}}(\mathbf{Tr}(\mathbf{D}(\Phi_k')), \nabla\Phi_i') - (\bar{v} \cdot \nabla\Phi_k' + \Phi_k' \cdot \nabla\bar{v}, \Phi_i') \\ D_i &= -\tfrac{1}{\mathrm{Re}} \int_{\Gamma_\sigma} \mathbf{Tr}(\mathbf{D}(\bar{v}) \cdot \nabla\Phi')dx - (\bar{v} \cdot \nabla\bar{v}, \Phi_i') \\ E_{ki} &= G \int_{\Gamma_\sigma} \Phi_k' \cdot \Phi_i' \\ K_i &= G \int_{\Gamma_\sigma} (\bar{v} - v^{cl})\Phi_i' dx \end{aligned}$$

5.5. Dynamic substructure/acoustic subdomain coupling

When studying the elastoacoustics of coupled fluid–structure systems modeled by the finite element method, it is clearly advantageous to reduce the size of the problem, since all of the DOFs of the fluid domain are added to those of the structure (we can, for example, cite the example of the interior of a car, which is likely to involve hundreds of thousands of DOFs). We need to reduce the number of unknowns to be processed without degrading the accuracy of the resulting solution.

However, when applying a modal synthesis method to a coupled vibroacoustic problem, we encounter two critical difficulties relating to the acoustic formulation and the choice of the associated subdomain method.

Indeed, even without considering applications of substructuring techniques, several finite element formulations have been proposed for the fluid problem when the fluid is coupled with a structure: formulations based on the pressure [ZIE 78], velocity potential [EVE 81], pressure/displacement potential [MOR 79], pressure/velocity potential [OLS 85] and the displacement [HAM 78, OLS 83, WAN 97]. Each of these methods has its own advantages and disadvantages. Pressure-based formulations lead to coupled vibroacoustic systems of the conventional form $A - w^2B$ (A and B, respectively, denote the stiffness and mass matrices, assembled over all DOFs of the structure + fluid) but whose matrices are not symmetric, which requires the application of numerically expensive transformations (irons transformations) to symmetrize the systems that we wish to solve. Formulations based on the velocity potential give rise to symmetric systems with the "unconventional" form $A - w^2B - iwC$ (C denotes the fluid–structure coupling matrix), requiring us to duplicate the solving space to obtain a system that can be used by efficient eigenvalue search algorithms. Formulations based on the pressure/displacement potential directly result in symmetric systems in the conventional form, but increase the number of DOFs at the nodes (these types of formulation also make it difficult to extend the local modal bases with static linking modes in the context of subdomain techniques). Finally, the special case of displacement-based formulations discretized by finite edge elements [BER 99] is extremely interesting from a numerical perspective, but is difficult to implement numerically, as the DOFs are not located at the nodes but on the edges of the elements.

Relatively few studies on the choice of the acoustic subdomain method are available in the literature, unlike for structural dynamics. This is because acoustic models are generally less expensive numerically (they usually require fewer DOFs per node, and the fluid mesh criteria are less restrictive than those associated with structures) and do not necessarily need to be reduced [WAN 97, BEN 99]. Ait Younes and Hamdi [AIT 97] proposed a subdomain method based on local modes with mixed interface conditions and a coupling with Lagrange multipliers (however, the numerical results that they obtain are purely numerical).

Vibroacoustic modal synthesis methods based on pressure formulations have now finally been proposed in order to couple fluid subdomains and substructures [XIN 96, SAN 01]. These methods require special transformations (to symmetrize the systems being solved) that are relatively delicate to implement with existing computation procedures.

The following section proposes a modal synthesis method for solving large-scale coupled fluid–structure problems. This method couples a dynamic substructuring method of the type proposed by Craig and Bampton with an acoustic subdomain method based on an acoustic formulation of the velocity potential. This choice is motivated, on the one hand, by the shape and symmetry properties of the coupled algebraic system thus obtained, and, on the other hand, by the fact that the local acoustic modal basis is easy to extend. Indeed, the choice of a formulation based on the velocity potential allows us to directly extend the method of Craig and Bampton to fluid domains. The local modes are then composed of modes with perfectly compliant interfaces extended by incompressible linking modes (analogously to the static linking modes in structural dynamics). The results obtained in the case of axisymmetric geometries decomposed into multiple fluid subdomains and multiple substructures demonstrate the validity and effectiveness of this method.

5.5.1. *Basic equations*

Consider a vibroacoustic problem divided into N_S substructures and N_F fluid subdomains. In the following, the exponents (and indices) s and f, respectively, index the substructures and fluid subdomains. Each substructure occupies a volume denoted as Ω_s. Each fluid subdomain occupies a volume Ω_f. There are three kinds of interface, defined as follows:

$$I^{ss'} = \Omega^s \cap \Omega^{s'}, \qquad J^{ff'} = \Omega^f \cap \Omega^{f'}, \qquad C^{sf} = \Omega^s \cap \Omega^f \qquad [5.54]$$

$I^{ss'}$ thus denotes the interface (or junction) between the substructure s and the substructure s' ($I^{ss'} = \emptyset$ if these two domains are not in contact). $J^{ff'}$ represents the fluid–fluid interface between the fluid subdomains f and f' ($\emptyset$ if there is no contact). C^{sf} is the fluid–structure interface between the substructure s and the fluid subdomain f ($\emptyset$ if Ω^s and Ω^f are not in contact).

5.5.1.1. *Equations for the structures*

We assume that each substructure is elastic, linear and isotropic, with no stress or initial strain. If there are no volume sources, the equation describing their vibrational behaviors is given by:

$$\nabla \cdot \sigma^s - \rho_s \ddot{\mathbf{u}}^s = 0, \qquad s = 1, \ldots, N_s \qquad [5.55]$$

$\rho_s, \mathbf{u}^s, \sigma^s$ are respectively the density, the displacement field and the stress tensor of the substructure s.

If we write Γ_u^s for the boundaries with imposed displacement and Γ_f^s for the boundaries with imposed external forces, the boundary conditions of the substructure s are given by:

$$\mathbf{u}^s|_{\Gamma_u^s} = \bar{\mathbf{u}}^s, \quad \sigma^s \cdot n|_{\Gamma_f^s} = \bar{f}^s \qquad [5.56]$$

At the substructure/substructure interfaces $I^{ss'}$, the displacement and the normal component of the stress tensor must be continuous. These conditions may be written as:

$$\left(\mathbf{u}^s - \mathbf{u}^{s'}\right)\Big|_{I^{ss'}} = 0 \qquad [5.57]$$

$$\left(\sigma^s \cdot n - \sigma^{s'} \cdot n\right)\Big|_{I^{ss'}} = 0 \qquad [5.58]$$

5.5.1.2. *Equations for fluids*

We consider small adiabatic perturbations of a perfect fluid around its rest state. We write $\mathbf{v}^f$, p^f, ρ_f and c_f for the acoustic velocity, the acoustic pressure, the density and the speed of sound in the subdomain f. Let φ^f be the acoustic velocity potential defined by:

$$\mathbf{v}^f = \frac{1}{\rho_f} \nabla \varphi^f \qquad [5.59]$$

The pressure and potential are related by the following equation:

$$p^f = -\frac{\partial \varphi^f}{\partial t} \qquad [5.60]$$

Moreover, the velocity potential satisfies the wave equation:

$$\frac{1}{\rho_f} \Delta \varphi^f - \frac{1}{\rho_f c_f^2} \ddot{\varphi}^f = 0, \qquad f = 1, \ldots, N_F \qquad [5.61]$$

We write Γ_φ^s and Γ_ν^s for the boundaries of the imposed potential (which also amounts to imposing the pressure according to equation [5.60] and the boundaries of the imposed normal acoustic velocity. The boundary conditions of the subdomain f are given by:

$$\varphi^f\big|_{\Gamma^f_\varphi} = \bar{\varphi}^f, \quad \left.\frac{\partial \varphi^f}{\partial n}\right|_{\Gamma^f_\nu} = \rho_f \bar{\nu}^f \qquad [5.62]$$

Note that a wall-type boundary condition with specific acoustic admittance β imposed on a surface Γ^f_β with normal n (directed outward from the fluid) would be written in the form of $(\partial\varphi^f/\partial n + \beta/c_f\dot{\varphi}^f)|_{\Gamma^f_\beta} = 0$. However, without loss of generality in the proposed method, we can assume in the following that there is no damping (either structural or acoustic).

The continuity conditions imposed on the fluid–fluid interfaces $J^{ff'}$ are based on the continuity of the pressure and normal velocity fields, i.e.:

$$\left.(\varphi^f - \varphi^{f'})\right|_{J^{ff'}} = 0 \qquad [5.63]$$

$$\left.\left(\frac{1}{\rho_f}\frac{\partial \varphi^f}{\partial n} - \frac{1}{\rho_{f'}}\frac{\partial \varphi^{f'}}{\partial n}\right)\right|_{J^{ff'}} = 0 \qquad [5.64]$$

5.5.1.3. *Coupling conditions at the fluid–structure interfaces*

If the substructure s and the fluid subdomain f are in contact, then the coupling conditions at the fluid–structure interface C^{sf} are given by:

$$\left.\frac{1}{\rho_f}\frac{\partial \varphi^f}{\partial n}\right|_{C^{sf}} = \dot{\mathbf{u}}^s \cdot n \qquad [5.65]$$

$$\sigma^s \cdot n|_{C^{sf}} = \dot{\varphi}^f n \qquad [5.66]$$

These conditions represent the continuity of the normal velocity and the continuity of the normal component of the stress tensor at the interface.

With the above notation, the boundaries $\partial\Omega^s$ (and $\partial\Omega^f$) of the substructure s (and the subdomain f) fully decompose as follows:

$$\partial\Omega^s = \Gamma_u^s \cup \Gamma_f^s \left(\bigcup_{\substack{s'=1 \\ s' \neq s}}^{N_S} I^{ss'} \right) \left(\bigcup_{f=1}^{N_F} C^{sf} \right),$$

$$\partial\Omega^f = \Gamma_\varphi^f \cup \Gamma_\nu^f \left(\bigcup_{\substack{f'=1 \\ f' \neq f}}^{N_F} J^{ff'} \right) \left(\bigcup_{s=1}^{N_S} C^{sf} \right) \qquad [5.67]$$

5.5.2. *Variational formulations*

5.5.2.1. *Variational formulation associated with the substructures*

Let $\mathbf{u}^{s^*}$ be an arbitrary test field associated with the substructure s. Equation [5.55] is integrated over the domain Ω^s. After integrating by parts then applying the conditions [5.56], [5.58] and [5.66], the variational problem consists of finding u^s such that $\left\{ \mathbf{u}^s|_{\Gamma_u^s} = \bar{\mathbf{u}}^s \right\}$ and:

$$\int_{\Omega^s} \rho_s \mathbf{u}^{s^*} \cdot \ddot{\mathbf{u}}^s dV + \int_{\Omega^s} \varepsilon^{s^*} : \sigma^s dV + \sum_{f=1}^{N_F} \int_{C^{sf}} \mathbf{u}^{s^*} \cdot n\dot{\varphi}^f dS$$

$$= \int_{\Gamma_f^s} \mathbf{u}^{s^*} \cdot \bar{f}^s dS + \sum_{\substack{s'=1 \\ s' \neq s}}^{N_s} \int_{I^{ss'}} \mathbf{u}^{s^*} \cdot (\sigma^{s'} \cdot n) dS \qquad [5.68]$$

for all $\mathbf{u}^{s^*}$ such that $\{\mathbf{u}^{s^*}|_{\Gamma_u^s} = 0\}$ and $s = 1, \ldots, N_s$; where ε denotes the strain tensor. n is the outward normal of Ω^s, except on the boundaries C^{sf}, where it points inward (outward from Ω^f).

5.5.2.2. *Variational formulation associated with the fluid subdomains*

Let φ^{f^*} be an arbitrary test field associated with the fluid subdomain f. Equation [5.61] is integrated over the domain Ω^f. After integrating by parts then applying the conditions [5.62], [5.64] and [5.65], we can show that the variational problem consists of finding φ^f such that $\left\{ \varphi^f|_{\Gamma_\varphi^f} = \bar{\varphi}^f \right\}$ and:

$$-\int_{\Omega^f} \frac{1}{\rho_f c_f^2} \varphi^{f^*} \ddot{\varphi}^f dV - \int_{\Omega^f} \frac{1}{\rho_f} \nabla \varphi^{f*} \cdot \nabla \varphi^f dV + \sum_{s=1}^{N_S} \int_{C^{sf}} \varphi^{f^*} \dot{\mathbf{u}}^s \cdot n dS$$

$$= -\int_{\Gamma_\nu^f} \varphi^{f^*} \bar{\nu}^f dS - \sum_{\substack{f'=1 \\ f' \neq f}}^{N_F} \int_{J^{ff'}} \varphi^{f^*} \frac{1}{\rho_{f'}} \frac{\partial \varphi^{f'}}{\partial n} dS$$

$$\forall \varphi^{f^*} / \{ \varphi^{f^*} \Big|_{\Gamma_\varphi^f} = 0 \} \quad f = 1, \ldots, N_f \quad [5.69]$$

where n is the outward normal of Ω^f.

5.5.3. *Discretization by finite elements*

Discretizing the structural variational problem (equation [5.68]) and the acoustic variational problem (equation [5.69]) leads to the following algebraic forms.

$$\langle u^{s^*} \rangle \left([M^s]\{\ddot{u}^s\} + [K^s]\{u^s\} + \sum_{f=1}^{N_F} \left[C^{sf}\right] \{\dot{\varphi}^f\} \right)$$

$$= \langle u^{s^*} \rangle \left(\{f^s\} + \sum_{\substack{s'=1 \\ s' \neq s}}^{N_S} \{f_I^{ss'}\} \right) \quad s = 1, \ldots, N_S \quad [5.70]$$

$$\langle \varphi^{f^*} \rangle \left(-[Q^f]\{\ddot{\varphi}^f\} - [H^f]\{\varphi^f\} + \sum_{s=1}^{N_S} \left[C^{sf}\right]^T \{\dot{u}^s\} \right)$$

$$- -\langle \varphi^{f^*} \rangle \left(\{\nu^f\} + \sum_{\substack{f'=1 \\ f' \neq f}}^{N_F} \{\nu_J^{ff'}\} \right) \quad f - 1, \ldots, N_F \quad [5.71]$$

The discretized terms correspond to the terms in formulations [5.68] and [5.69] in the order that they arise. The notations $\langle \cdot \rangle$ and $\{\cdot\}$ denote row and column vectors, respectively.

$\{u^s\}$ and $\{\varphi^f\}$ contain all of the unknown DOFs associated with the structural displacement and the acoustic potential, respectively (the DOFs on the boundaries Γ_u^s and Γ_φ^s, which are known, are not included in these vectors). $[M^s]$, $[K^s]$ and $\{f^s\}$, respectively, denote the mass matrix of the substructure s, its stiffness matrix and its vector of equivalent external forces (including excitations inherent to the boundary Γ_u^s, with its imposed displacement). $[Q^f]$, $[H^f]$ and $\{\nu^f\}$ represent the mass matrix of the subdomain f, its stiffness matrix and its vector of external acoustic excitations (including excitations inherent to Γ_φ^s). The (ambiguous) terms of acoustic "mass" and "stiffness" are associated with the potential and kinetic acoustic energy, respectively (the opposite is true for the structures).

The term $\{f_t^{ss'}\}$ physically represents the interfacial forces exerted on the substructure s by the adjacent substructure s'. Similarly, the term $\{\nu_f^{ff'}\}$ represents the interfacial actions exerted on f by the adjacent fluid subdomain f'. The effects of these terms only apply to the DOFs of the corresponding interfaces.

We assemble the N_S substructures and the N_F acoustic subdomains into a global vector containing all of the DOFs of both the fluid and the structure, organized as follows:

$$\langle u \rangle = \langle u^1 \; u^2 \; \dots \; u^{N_s} \mid \varphi^1 \; \varphi^2 \; \dots \; \varphi^{N_F} \rangle \qquad [5.72]$$

Now, noting that the test vector $< u^* >$ is arbitrary, we can show that assembling the formulations [5.70] and [5.71] leads to the following algebraic system:

$$[M]\{\ddot{u}\} + [C]\{\dot{u}\} + [K]\{u\} = \{f\} + \{f_I\} \qquad [5.73]$$

where

$$[M] = \begin{bmatrix} M^1 & 0 & 0 & 0 \\ & \ddots & & \\ 0 & M^{N_s} & 0 & 0 \\ 0 & 0 & -Q^1 & 0 \\ & & \ddots & \\ 0 & 0 & 0 & -Q^{N_F} \end{bmatrix}$$

$$[K] = \begin{bmatrix} K^1 & 0 & 0 & 0 \\ & \ddots & & \\ 0 & K^{N_s} & 0 & 0 \\ 0 & 0 & -H^1 & 0 \\ & & \ddots & \\ 0 & 0 & 0 & -H^{N_F} \end{bmatrix}$$

$$[C] = \begin{bmatrix} 0 & 0 & C^{11} & \dots & C^{1N_F} \\ & \ddots & \vdots & & \vdots \\ 0 & 0 & C^{N_S 1} & \dots & C^{N_S N_F} \\ & & 0 & & 0 \\ & sym & & \ddots & \\ & & 0 & & 0 \end{bmatrix}$$

$$\{f\} = \left\{ \begin{array}{c} f^1 \\ \vdots \\ f^{N_S} \\ -\nu^1 \\ \vdots \\ -\nu^{N_F} \end{array} \right\} \quad \text{and} \quad \{f_I\} = \left\{ \begin{array}{c} \sum\limits_{s' \neq 1} f_I^{1s'} \\ \vdots \\ \sum\limits_{s' \neq N_S} f_I^{N_S s'} \\ -\sum\limits_{f' \neq 1} \nu_J^{1f'} \\ \vdots \\ -\sum\limits_{f' \neq N_F} \nu_J^{N_F f'} \end{array} \right\}$$

In the above expressions, the matrices $[C^{sf}]$ are implicitly zero when there is no interface between Ω^s and Ω^f $(C^{sf} = \emptyset)$.

The matrices $[M]$, $[C]$ and $[K]$ are symmetric. This property is essential, since it allows us to obtain coupled natural modes that satisfy orthogonality conditions. The final term of equation [5.73] will later vanish when we require continuity in the displacements between substructures and in the pressures between fluid subdomains.

For large-scale vibroacoustic problems, solving the global assembled system [5.73] can become highly expensive in terms of memory and computation time, and may even be impossible.

5.5.4. *Calculating the local modes*

Local modes of the substructures

The vector of DOFs of each substructure s is partitioned into the internal DOFs (indexed by i) and the DOFs of the junctions (indexed by j). The latter are the DOFs located at the interfaces between the substructure s and all other adjacent substructures. We can therefore write:

$$\langle u^s \rangle = \langle u_i^s \quad u_j^s \rangle, \quad [M^s] = \begin{bmatrix} M_{ii}^s & M_{ij}^s \\ M_{ji}^s & M_{jj}^s \end{bmatrix}, \quad [K^s] = \begin{bmatrix} K_{ii}^s & K_{ij}^s \\ K_{ji}^s & K_{jj}^s \end{bmatrix} \qquad [5.74]$$

Following Craig and Bampton's method, the selected local modes are those of the fixed interfaces $I^{ss'}$. They satisfy the following boundary value problem:

$$[K_{ii}^s - \omega^2 M_{ii}^s]\{\psi_i^s\} = 0, \qquad s = 1, \ldots, N_S \qquad [5.75]$$

Note that the fluid–structure interfaces C^{sf} are implicitly assumed to be free (i.e. the adjacent fluids do not act on the substructure).

These modes, which are orthogonal, are extended by static linking modes, which are defined as the static deformation of the substructure after applying a unit displacement to each of the DOFs at the junction, while forcing all others to be 0.

The local modal basis of a substructure is thus given by:

$$[\Psi^s] = \begin{bmatrix} \psi^s & -K_{ii}^{s\,-1} K_{ij}^s \\ 0 & I_{jj}^s \end{bmatrix} \qquad [5.76]$$

where $[\psi^s]$ is the matrix of fixed-interface modes arranged as columns. The physical DOFs of each substructure may then be decomposed according to their respective local modal bases, so that:

$$\{u^s\} = [\Psi^s]\{\alpha^s\}, \qquad s = 1, \ldots, N_S \qquad [5.77]$$

where $\{\alpha^s\}$ is the generalized coordinate vector of the substructure s, which contains both the coefficients of the fixed-interface modes and the physical DOFs (nodal displacements, rotations, etc.) of the structural junction.

Local modes of the acoustic subdomains

The form of the algebraic fluid system, which is inherent to the velocity potential formulation, allows us to directly extend Craig and Bampton's method. After taking a partition analogous to the one chosen for the substructures, we can write:

$$\langle \varphi^f \rangle = \langle \varphi_i^f \quad \varphi_j^f \rangle, \quad [Q^f] = \begin{bmatrix} Q_{ii}^f & Q_{ij}^f \\ Q_{ji}^f & Q_{jj}^f \end{bmatrix}, \quad [H^f] = \begin{bmatrix} H_{ii}^f & H_{ij}^f \\ H_{ji}^f & H_{jj}^f \end{bmatrix} \qquad [5.78]$$

We then define the local modes with perfectly compliant interfaces $J^{ff'}$ (zero pressure) by:

$$[H_{ii}^f - \omega^2 Q_{ii}^f]\{\phi_i^f\} = 0, \qquad f = 1, \ldots, N_F \qquad [5.79]$$

where the fluid–structure interfaces C^{sf} are implicitly assumed to be perfectly rigid (i.e. the adjacent substructures do not act on the fluid subdomain).

This basis is extended by direct analogy with the static linking modes from structural dynamics. In acoustics, these modes physically correspond to incompressible linking modes ($C_f \to \infty$ is equivalent to $\omega \to 0$). They are defined as the incompressible response of the subdomain after applying a unit potential to each of the DOFs of the fluid junction, while forcing all others to be 0.

Finally, the local modal basis of the fluid subdomain f is given by:

$$[\Phi^f] = \begin{bmatrix} \phi^f & -H_{ii}^{f^{-1}} H_{ij}^f \\ 0 & I_{jj}^f \end{bmatrix} \qquad [5.80]$$

where $[\phi^f]$ represents the matrix of modes with compliant interfaces arranged as columns. The physical DOFs of each fluid subdomain may now be decomposed according to their respective local modal bases, so that:

$$\{\varphi^f\} = [\Phi^f]\{\beta^f\}, \qquad f = 1, \ldots, N_F \qquad [5.81]$$

where $\{\beta^s\}$ is the generalized coordinate vector of the subdomain f, which contains both the coefficients of the modes with compliant interfaces and the physical DOFs (i.e. the nodal potentials) of the fluid junction.

Finally, we should note that directly extending Craig and Bampton's method in acoustic problems only seems to be possible for pressure-based or potential-based formulations. The forms of the other types of acoustic formulation prevent us from considering *a priori* the static case ($\omega = 0$) or the incompressible case when we derive the nodes for the extension.

5.5.5. *Modal synthesis*

Model reduction

The local decompositions [5.77] and [5.81] can be assembled as follows:

$$\{u\} = [\Psi]\{p\} \qquad [5.82]$$

with

$$[\Psi] = \begin{bmatrix} \Psi^1 & 0 & 0 & 0 \\ & \ddots & & \\ 0 & \Psi^{N_s} & 0 & 0 \\ 0 & 0 & \Phi^1 & 0 \\ & & \ddots & \\ 0 & 0 & 0 & \Phi^{N_F} \end{bmatrix}, \quad \{p\} = \begin{Bmatrix} \alpha^1 \\ \vdots \\ \alpha^{N_S} \\ \beta^1 \\ \vdots \\ \beta^{N_F} \end{Bmatrix} \qquad [5.83]$$

Then, after projecting, equation [5.73] becomes:

$$[M_p]\{\ddot{p}\} + [C_p]\{\dot{p}\} + [K_p]\{p\} = \{f_p\} + [\Psi]^T\{f_I\} \qquad [5.84]$$

with $[M_p] = [\Psi]^T[M][\Psi]$, $[C_p] = [\Psi]^T[C][\Psi]$, $[K_p] = [\Psi]^T[K][\Psi]$ and $\{f_p\} = [\Psi]^T\{f\}$.

We must now consider the continuity conditions at the structure–structure and fluid–fluid interfaces. Indeed, the DOFs of $\{p\}$ are not linearly independent. The linear relationships between these DOFs may be derived from the equality of the displacements at structure–structure interfaces and the equality of the pressure at fluid–fluid interfaces. They can be expressed by a global connectivity matrix $[S]$:

$$\{p\} = [S]\{q\} \qquad [5.85]$$

where $\{q\}$ only contains the linearly independent DOFs. $[S]$ characterizes both the connectivity between substructures and the connectivity between fluid subdomains. For methods such as the Craig–Bampton type method considered here, the matrix $[S]$ is Boolean and has a straightforward expression, since the physical DOFs of the junction are explicitly included in the generalized unknowns $\{p\}$.

From the conditions [5.58] and [5.64], we have the following compatibility equations:

$$\left\{f_I^{ss'}\right\} + \left\{f_I^{s's}\right\} = 0, \qquad \left\{\nu_J^{ff'}\right\} + \left\{\nu_J^{f'f}\right\} = 0 \qquad [5.86]$$

We can now show that these equations imply that $[S]^T[\Psi]^T\{f_I\} = [S]^T\{f_I\} = 0$ (for more details, refer, for example, to [CRA 95]). Thus, the final system to be solved is:

$$[M_q]\{\ddot{q}\} + [C_q]\{\dot{q}\} + [K_q]\{q\} = \{f_q\} \qquad [5.87]$$

with

$$\begin{aligned} [M_q] &= [S]^T[M_p][S],\ [C_q] = [S]^T[C_p][S],\ [K_q] = [S]^T[K_p][S], \\ \{f_q\} &= [S]^T\{f_p\} \end{aligned} \qquad [5.88]$$

In practice, this is a strongly reduced model relative to system [5.73], since its size is equal to the total number of orthogonal local modes after truncation plus the total number of DOFs of the junction.

Obtaining the coupled natural modes

In order to be able to apply effective eigenvalue search algorithms, we must first put the system into the following equivalent form:

$$[A]\{y\} + [B]\{\dot{y}\} = \{F\} \qquad [5.89]$$

where

$$\{y\} = \begin{Bmatrix} q \\ \dot{q} \end{Bmatrix},\ [A] = \begin{bmatrix} K_q & 0 \\ 0 & -M_q \end{bmatrix},\ [B] = \begin{bmatrix} C_q & M_q \\ M_q & 0 \end{bmatrix},$$

$$\{F\} = \begin{Bmatrix} f_q \\ 0 \end{Bmatrix} \qquad [5.90]$$

This artificially doubles the solution space. The two matrices $[A]$ and $[B]$ are real and symmetric.

We set a time dependency of the form $\{y\} = \{y_m\}e^{-\mathrm{i}\omega_m t}$. The global modes of the coupled fluid–structure system are then obtained by solving the following global eigenvalue problem:

$$([A] - \mathrm{i}\omega_m[B])\{y_m\} = \{0\} \qquad [5.91]$$

Solving this problem leads to pairs of angular frequencies associated with complex conjugate eigenvectors. Their orthogonality properties are preserved as a result of the symmetry of the matrices. If we choose to normalize relative to the mass matrix of the coupled system, given by $\langle\bar{q}_n\rangle[M_q]\{q_n\} = 1$, we can show that these properties may be written as:

$$\langle\bar{y}_n\rangle[B]\{y_m\} = a_m\delta_{nm}, \qquad \langle\bar{y}_n\rangle[A]\{y_m\} = \omega_m a_m\delta_{nm} \qquad [5.92]$$

where $a_m = \langle\bar{q}_m\rangle[C_q]\{q_m\} + 2\mathrm{Im}(\omega_m)$, δ_{nm} denotes the Kronecker delta and the bar denotes (complex) conjugation. Note that the imaginary parts of the angular frequencies are zero because we neglected the damping.

If we wish to take damping into account (either structural or acoustic damping), the method presented in this chapter and the associated solving techniques remain the same: in this case, the matrix $[C]$ would include both coupling and damping effects. The only difference is that the angular frequencies thus obtained would have non-zero (negative) imaginary parts.

Obtaining the vibroacoustic response

The vibroacoustic response that we wish to find has now been decomposed over the coupled modal basis obtained above:

$$\{y\} = [Y]\{\alpha\} \qquad [5.93]$$

where the matrix $[Y]$ is the matrix of modes $\{y_m\}$ arranged as columns ($m = 1, \ldots, M$ where M is the number of coupled modes left after truncation). The orthogonality properties [5.92] allow us to diagonalize the system [5.89], namely by setting $h_m = a_m^{-1}\langle \bar{y}_m \rangle \{F\}$:

$$\dot{\alpha}_m + \mathrm{i}\omega_m \alpha_m = h_m \qquad [5.94]$$

The solution of this equation in analytical form is given by:

$$\alpha_m(t) = \int_0^t e^{-\mathrm{i}\omega_m(1-\tau)} h_m(\tau) d\tau + \alpha_m(0) e^{-\mathrm{i}\omega_m t} \qquad [5.95]$$

In the steady state, we can take the Fourier transform of [5.94], leading to the following frequency response:

$$\alpha_m(\omega) = \frac{h_m(\omega)}{\mathrm{i}(\omega_m - \omega)} \qquad [5.96]$$

The physical response is then reconstructed by successively taking the transformations [5.82], [5.85] and [5.94].

5.6. Numerical simulation

Two examples of fluid–structure interaction are discussed below. The first is a two-dimensional elastic circular ring, for which we will separately validate the inertial effects of the fluid, the method of acoustic subdomains, the substructuring method and substructures coupled with the acoustic fluid. The second example is a two-dimensional boat propeller decomposed into four substructures and coupled with a fluid domain. We will present the results of the model in both the dry and immersed cases.

5.6.1. *Elastic ring*

The objective is to validate the proposed model. A two-dimensional model with axisymmetric geometry is established. A ring structure is coupled with a closed volume of fluid filled with water to simulate strong coupling between the structure and the fluid. We will calculate the bending and tension modes of an elastic ring of radius R, thickness h, density μ and Young's modulus E contained in a circular cavity of radius R' and containing a fluid of density

ρ. The matrices of the finite element system are generated using Ansys code, with the fluid–structure coupling scheme. The finite element model consists of quadrilateral elements for the fluid and the structure in Ansys, namely the FLUID29 and PLANE42 elements. The bold circle represents the elastic ring, as shown in Figure 5.2.

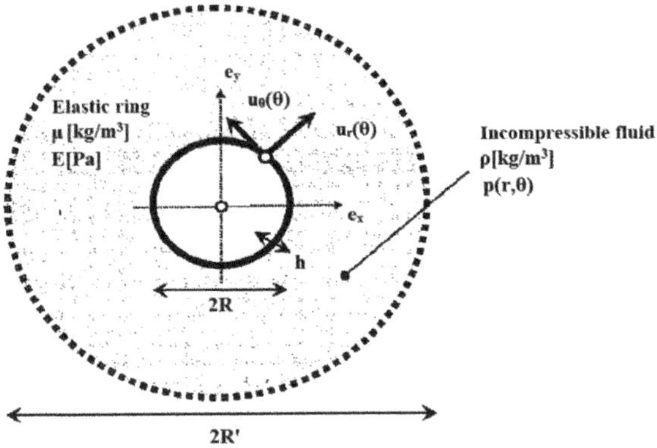

Figure 5.2. *Elastic ring coupled with an incompressible fluid contained in a cylindrical cavity*

The equations of the coupled problem are given below [SIG 11]:

– *Structure*: The state of the structure is described by the radial displacement $u_r(\theta)$ and the orthoradial displacement $u_\theta(\theta)$, which satisfy the equations:

$$-\omega^2 \mu h u_r + \frac{Eh}{R^2}\left[u_r + \frac{\partial u_\theta}{\partial \theta} + \frac{h^2}{12R^2}\left(\frac{\partial^4 u_r}{\partial \theta^4} - \frac{\partial^3 u_\theta}{\partial \theta^3}\right)\right]$$

$$= -p|_{r=R} \quad \theta \in [0, 2\pi]$$

$$-\omega^2 \mu h u_\theta - \frac{Eh}{R^2}\left[\frac{\partial^2 u_\theta}{\partial \theta^2} + \frac{\partial u_r}{\partial \theta}\frac{h^2}{12R^2}\left(\frac{\partial^2 u_\theta}{\partial \theta^2} - \frac{\partial^3 u_r}{\partial \theta^3}\right)\right] = 0$$

$$\theta \in [0, 2\pi]$$

– *Fluid*: The state of the fluid is described by the pressure $p(r, \theta)$, which satisfies the equation:

$$\frac{\partial^2 p}{\partial r^2} + \frac{1}{r}\frac{\partial p}{\partial r} + \frac{1}{r^2}\frac{\partial^2 p}{\partial \theta^2} = 0 \quad \text{for} \quad (\mathrm{r}, \theta) \in [\mathrm{R}, \mathrm{R}'] \times [0, 2\pi] \qquad [5.97]$$

with the boundary conditions:

$$\left.\frac{\partial p}{\partial r}\right|_{r=R} = \rho\omega^2 u_r \quad p|_{r=R'} = 0 \tag{5.98}$$

It is possible to write down an analytical solution to the problem as stated, based on the Fourier series expansion of the unknowns u_r, u_θ and p:

$$u_r(\theta) = u_r^0 + \sum_{m\geq 1} u_r^m \cos(m\theta) + \sum_{m\geq 1} u_r^{m'} \sin(m'\theta)$$

$$u_r(\theta) = \sum_{m\geq 1} u_\theta^m \sin(m\theta) + \sum_{m'\geq 1} u_\theta^{m'} \cos(m'\theta)$$

and:

$$p(r,\theta) = p_0 + \sum_{m\geq 1} p_m^m \cos(m\theta) + \sum_{m'\geq 1} u_r^{m'} \sin(m'\theta)$$

Without loss of generality, we will restrict attention to the symmetric Fourier components ($m = 0$ and $m \geq 1$) ; for these components, the above equations become [SIG 11]:

$$-\omega^2\mu h u_r^m + \frac{Eh}{R^2}\left[u_r^m + m u_\theta^m + \frac{h^2}{12R^2}\left(m^4 u_r^m + m^3 u_\theta^m\right)\right] = -p_m|_{r=R}$$

$$-\omega^2\mu h u_\theta^m - \frac{Eh}{R^2}\left[m^2 u_\theta^m + m u_r^m \frac{h^2}{12R^2}\left(m^2 u_\theta^m + m^3 u_r^m\right)\right] = 0$$

where the pressure field p_m satisfies:

$$\frac{\partial^2 p_m}{\partial r^2} + \frac{1}{r}\frac{\partial p_m}{\partial r} - \frac{m^2}{r^2} p_m = 0 \tag{5.99}$$

with

$$\left.\frac{\partial p_m}{\partial r}\right|_{r=R} = \rho\omega^2 u_r^m \qquad p_m|_{r=R'} = 0 \tag{5.100}$$

It can be shown that, for $m \neq 0$, the general form of the solution to the above equation is $p^m(r) = \alpha_m r^m + \frac{\beta^m}{r^m}$, where the constants α_m and β_m are determined by the boundary conditions; we find that:

$$\begin{cases} m(\alpha_m R^{m-1} - \frac{\beta_m}{R^{m+1}}) = \rho\omega^2 u_r^m \\ (\alpha_m R'^m + \frac{\beta_m}{R'^m}) = 0 \end{cases} \quad [5.101]$$

which implies that

$$\alpha_m = +\frac{1}{m}\frac{R^{m+1}}{R'^{2m} + R^{2m}}\rho\omega^2 u_r^m \qquad \beta_m = -\frac{R'^{2m}}{m}\frac{R^{m+1}}{R'^{2m} + R^{2m}}\rho\omega^2 u_r^m$$

[5.102]

The pressure field is:

$$p_m(r) = +\rho\frac{1}{m}\frac{R^{m+1}}{R'^{2m} + R^{2m}}(r^m - \frac{R'^{2m}}{r^m})\omega^2 u_r^m \quad [5.103]$$

and we can now calculate the pressure at the wall from:

$$p_m|_{r=R} = -\rho\frac{R}{m}\frac{R'^{2m} - R^{2m}}{R'^{2m} + R^{2m}}\omega^2 u_r^m \quad [5.104]$$

Substituting into the equation of motion of the structure, we have that:

$$[-\omega^2(M_S^m + M_A^m) + K_S^m]U_m = 0 \quad [5.105]$$

with $U_m^T =< u_r^m, u_\theta^m >$, and the added mass matrix M_A^m is given by:

$$M_A^m = \begin{bmatrix} \rho R\mu^m & 0 \\ 0 & 0 \end{bmatrix} \quad [5.106]$$

where the added mass coefficient μ_m is defined as:

$$\mu_m = \frac{1}{m}\frac{R'^{2m} - R^{2m}}{R'^{2m} + R^{2m}} \quad [5.107]$$

Modal analysis

The geometric and physical parameters of the stated problem are listed in Table 5.1. The calculation was performed with $I = 192$ finite fluid elements.

Structure parameters	Young's E mod. (GPa)	Density μ (kg/m^3)	h (m)	R (m)
	210	7,800	0.05	0.5
Fluid parameters	–	Density ρ (kg/m^3)	R' (m)	R (m)
	–	1000	0.3	0.5

Table 5.1. *Geometric and physical properties*

For the Fourier components $m = 2$ to $m = 5$, Figure 5.3 shows a representation of the modal shapes of the ring added to the fluid pressure field, deduced from equation [SIG 11]: $P_m = -\rho\omega^2 K_F^{-1} R^T U_m$.

Table 5.2 characterizes the inertial effects using the coefficient $\beta = f_{\text{with fluid}}/f_{\text{without fluid}}$, which is the ratio of the natural frequencies of the modes with and without the fluid for the Fourier components $m = 0$ to $m = 5$.

For the components with orders $m = 0$ and $m = 1$, the natural modes, respectively, correspond to rotation and translation of the ring: these are rigid body modes that do not induce a motion in the fluid.

Fourier components	β
$m = 0$	–
$m = 1$	–
$m = 2$	67.5%
$m = 3$	69.8%
$m = 4$	73.1%
$m = 5$	75.4%

Table 5.2. *Characterization of the inertial effects*

The inertial effects become less and less visible as the order of the modes increases; this can be highlighted quantitatively by calculating the added mass and qualitatively by observing the shape of the pressure field, whose fluctuations are increasingly localized around the ring. The higher order modes generate local motion in the fluid: the associated kinetic energy is lower than the global motion, which results in a less significant inertial effect.

Table 5.3 compares the analytical calculations [SIG 11] and numerical computations performed by Ansys for the natural frequencies of the immersed elastic ring.

Natural frequencies (Hz)	Analytical solution	Numerical computation
F_3	104	103.70
F_4	307	304.36
F_5	607	613.55
F_6	1,004	1,029.0

Table 5.3. *Analytical and numerical calculations of the natural frequencies for the immersed elastic ring*

Figures 5.4 and 5.5 show the composite substructures and subdomains of this fluid–structure interaction substructuring problem.

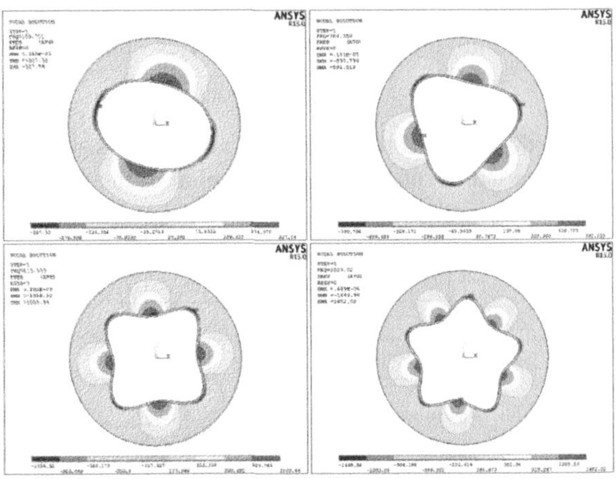

Figure 5.3. *Modal shapes of the elastic ring coupled with an incompressible fluid, Fourier components $m = 2$ to $m = 5$. For a color version of this figure, see www.iste.co.uk/elhami/interactions.zip*

Decomposition of the circular acoustic cavity

The sole purpose of this test case is to validate the subdomain method in acoustics, without coupling. The example is that of a circular fluid cavity divided into two and four subdomains (Figures 5.4(b) and 5.4(c)). The walls of the cavity are perfectly rigid. Table 5.4 indicates the natural frequencies obtained for both the full model and the subdomains.

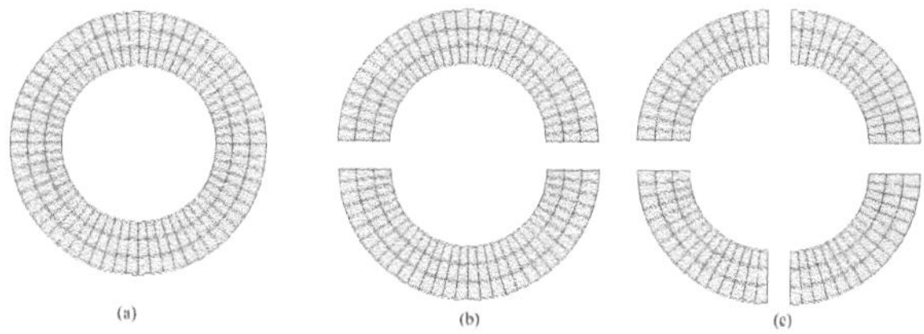

Figure 5.4. *Finite element mesh of the subdomains. For a color version of this figure, see www.iste.co.uk/elhami/interactions.zip*

Natural frequencies (Hz)	Full model (a)	Subdomain (b)	Subdomain (c)
1	–	–	–
2	603.36	603.36 (0.0%)	603.39 (0.0%)
3	1,204.4	1,204.4 (0.0%)	1,204.4 (0.0%)
4	1,801.1	1,801.1 (0.0%)	1,802.0 (0.04%)
5	2,391.5	2,391.5 (0.0%)	2,391.5 (0.0%)
6	2,974.7	2,974.7 (0.0%)	2,979.1 (0.14%)

Table 5.4. *Natural frequencies of the circular cavity*

The results are in good agreement, and the errors $\Delta f_{\text{rel}} = (f_{(b,c)} - f_{(a)})/f_{(a)}$ relative to the reference frequency $f_{(a)}$ of the full model are shown. They are satisfactory, and do not exceed 0.2%.

Decomposing the elastic ring

Table 5.5 lists the natural frequencies of the dry structure (without coupling with the surrounding fluid). The ring is also divided into two and four substructures of equal length, which means that there are two interfaces for each substructure (Figure 5.5).

Table 5.6 gives the natural frequencies of the coupled fluid/structure system (Figure 5.6). In this case, solving this requires us to duplicate the solution space, as mentioned above. The results show the validity of the method, as the values of the errors are all less than 0.2%.

Table 5.6 lists the natural frequencies of the dry structure (obtained by direct calculation). We can clearly see the effect of coupling in the fact that the natural frequencies are significantly lower. The correlation between the results obtained by direct calculation and substructuring demonstrates the validity and the effectiveness of the proposed vibroacoustic substructuring method.

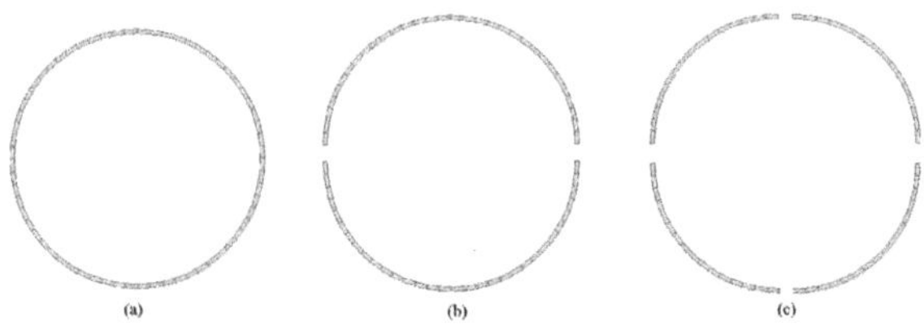

Figure 5.5. *Finite element mesh of the substructures. For a color version of this figure, see www.iste.co.uk/elhami/interactions.zip*

Natural frequencies (Hz)	Full model (a)	Substructure (b)	Substructure (c)
1	–	–	–
2	–	–	–
3	153.51	153.51 (0.0%)	153.51 (0.0%)
4	435.47	435.51 (0.01%)	435.47 (0.0%)
5	838.54	838.61 (0.01%)	838.54 (0.0%)
6	1,363.6	1,364.3 (0.05%)	1,363.7 (0.01%)

Table 5.5. *Natural frequencies of the elastic ring*

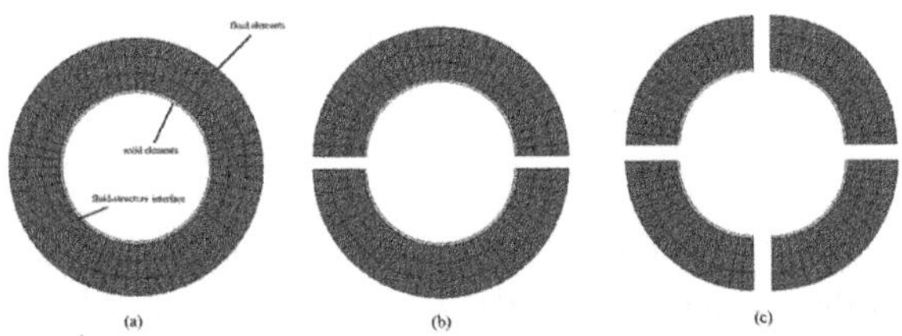

Figure 5.6. *Finite element mesh of the immersed ring. For a color version of this figure, see www.iste.co.uk/elhami/interactions.zip*

5.6.2. *Boat propeller*

We study the dynamic behavior of a boat propeller. The geometric model of this propeller (Figure 5.7), as well as the mesh and geometric substructuring (Figure 5.8), was performed using Ansys.

Natural frequencies (Hz)	Full model (a)	Substructure (b)	Substructure (c)
1	–	–	–
2	–	–	–
3	103.70	103.70 (0.0%)	103.70 (0.0%)
4	304.36	304.25 (0.03%)	303.71 (0.2%)
5	613.55	612.04 (0.2%)	613.73 (0.03%)
6	1,029.0	1,038.2 (0.9%)	1,021.7 (0.7%)

Table 5.6. *Natural frequencies of the immersed ring*

Figure 5.7. *Boat propeller. For a color version of this figure, see www.iste.co.uk/elhami/interactions.zip*

The mesh was constructed using quadrilateral elements. To perform the modal synthesis calculation (reduction of DOF), we divide the propeller into four substructures and present the results obtained for the full model and for the four substructures, taking into account the fluid–structure interaction. The material properties of this problem are listed in Table 5.7.

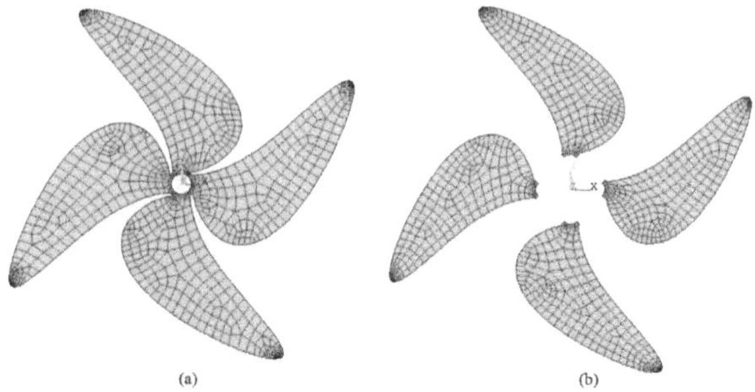

Figure 5.8. *Finite element mesh and substructures. For a color version of this figure, see www.iste.co.uk/elhami/interactions.zip*

Structure	E (GPa)	ρ_s (kg/m^3)	Poisson coefficient ν
	210	$7,860$	0.3
Fluid	Speed of sound c (m/s)	ρ_f (kg/m^3)	–
	$1,500$	$1,000$	–

Table 5.7. *Material properties*

In Table 5.8, the modal analysis of the propeller is presented, and the calculated natural frequencies are compared (between the cases of the dry and immersed propeller, on the one hand, and in terms of the substructures, on the other hand). The natural modes are shown in Figure 5.9.

Dry propeller	Immersed propeller	
(Hz)	Full model (a)	Substructure (b)
107.09	68.429	67.147 (0.01%)
198.03	96.405	96.916 (0.0%)
298.05	127.06	130.25 (0.02%)
422.94	141.31	139.94 (0.01%)

Table 5.8. *Natural frequencies of the propeller*

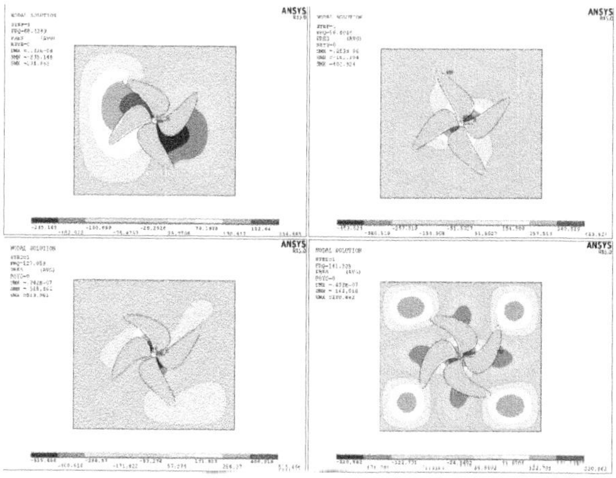

Figure 5.9. *Natural modes of the immersed propeller. For a color version of this figure, see www.iste.co.uk/elhami/interactions.zip*

6

Reliability-based Optimization for FSI

6.1. Introduction

Guaranteeing the reliability of components subject to progressive deterioration and improving our understanding of their mechanisms with the objective of optimizing their usage and reducing the risk of failure while considering the added factor of cost efficiency is a major challenge. Reliability theory uses probabilistic estimates as a measure. The question of whether a given probability value is acceptable is highly complex, and it is clear that any decision must consider the quality of the available information, the estimated risk level, the consequences of failure in terms of material damage and personal injury and the operating lifetime of the risk with the understanding that some risks are imposed on everyone by their living conditions, and others are voluntary.

As discussed in Chapter 5, optimization aims to automatically determine the best design with respect to criteria describing the structural performance. The solution constructed by this iterative process, which alternates between structural analysis and optimization techniques, is called the optimal design. However, optimizing the structure without considering reliability can result in high risks of failure relative to the designer's target failure probability threshold. The objective of reliability-based optimization is to propose designs that guarantee the desired level of reliability.

In this chapter, we begin by giving a detailed presentation of the concept of reliability, its objectives and its advantages in mechanics. We will continue by presenting its implementation in the context of reliability-based design optimization (RBDO) problems for structures subject to vibrations that interact with flowing fluids with the objective of identifying the critical frequency bands that might cause the structure being optimized to experience damage or destruction.

6.2. Reliability in mechanics

Reliability is the discipline that studies the failure risk of arbitrary devices, and is in some sense the science of failure. It relies on mathematical theories that are constantly being perfected, technological understanding and experience.

Indeed, reliability analysis was the unambiguous result of introducing statistical and probabilistic theories, and it is based on experience. It fulfills a key, indispensable role in operational safety studies. Originally, reliability was studied in connection with high-technology systems (nuclear centers and the aerospace sector). Today, reliability has become a key quality and decision-making parameter in the study of most components, products and processes.

Many industrial actors actively work to assess and improve the reliability of their products as part of the development cycle, from the design phase to commissioning (design, manufacture and operation), with the goal of expanding their knowledge of the cost/reliability ratio and gaining control over sources of failure.

Reliability analysis in mechanics is a very important tool for characterizing the behavior of products in each phase of their lifecycle, measuring the impact of design modifications on product integrity, labeling new products and continuously improving product performance.

In structural reliability analysis, the parameters of influence are viewed as random variables, and the probability of failure is calculated from a physical failure equation whose variables are formulated with probabilistic expressions.

6.2.1. *Random variables*

The design of a structure or a mechanical component needs to satisfy certain requirements. By working from an understanding of these requirements and using appropriate methods, the design of the structure can be made to account for external interactions and material properties.

These methods are derived from theories that approximate reality. Each different branch of physics gives a model of the phenomena that it considers, and these models are typically implemented numerically. When passing from reality to the physical model, and from the physical model to the numerical model, some parameters are neglected for purposes of simplification or due to ignorance.

These model errors often affect the results and can only be reduced by using more accurate models.

The structure is then subjected to external actions or unknowns, which are estimated from a finite set of measurements. Some parameters vary over space and time, such as the ocean swell, the wind or the traffic on a bridge. These kinds of parameters are difficult to determine precisely and therefore are difficult to model.

The construction quality of the structure, the chosen materials and their capacity to respond to external actions are not always perfectly known. Indeed, material properties are determined based on laboratory or field tests and statistical procedures. Depending on the quality of these procedures, the representativeness of the tests, manufacturing precision and the range of values that these parameters take can vary. These parameters are examples of unknowns that are said to be internal to the structure (Figure 6.1). Measuring errors can be reduced by improving the accuracy of measurements and reducing the statistical errors that arise from the accumulation of data. Some uncertainties relating to the load and the material properties form part of the inherent variability of the physical phenomenon. These uncertainties are irreducible unless the phenomenon itself is modified.

Ambiguous or approximate measurements of material properties or of the various parameters can lead to inaccuracies that contribute to the unknowns. Furthermore, the theories used, which are not always fully developed, and the chosen parameters, which cannot always be determined at arbitrary times t in some experimental contexts, also give rise to uncertainty, which also generates unknowns. The influence of these unknowns on the capacity of the structure to respond to the initial requirements can be both positive and negative. To control them, we need reliable methods.

In the context of the reliability of mechanical structures, we are interested in finding the probability that the structure will be capable of meeting all of the requirements for which it was designed, without failure. This probability is defined as the failure or non-execution value at a given time for a given clause or commitment.

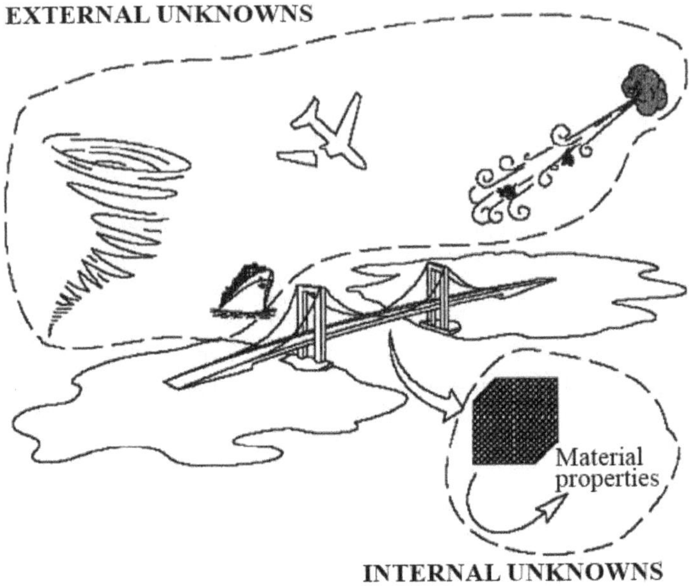

Figure 6.1. *Internal and external unknowns*

If the unknown is time dependent, such as the wind or the ocean swell, it can be modeled by a stochastic process. If it is time independent, like most geometric dimensions, it can be represented by a random variable. If the unknown varies over space, its representation can be based on a random field. Depending on the type of unknown, some methods are more applicable than others. For stochastic processes, other mathematical models must be used.

Applying stochastic processes, random variables and random fields allows us to begin to master the uncertainty, but does not solve every problem that we might encounter.

6.2.2. *Reliability function*

The reliability of a device after a time t is given by the probability that this device does not experience a failure between 0 and t. Denoting by t the random variable characterizing the time at which the device fails, the reliability may be described by the function $R(t)$ (reliability) such that:

$R(t)$ = Prob (that an entity E does not fail in the interval $[0,t]$, given that it is not defective at time $t = 0$)

$$R(t) = 1 - F(t) \qquad [6.1]$$

$F(t)$ is the distribution function of the variable t.

Note that the "time" variable should be viewed in units corresponding to the usage period.

In general, the reliability function has the form given in Figure 6.2.

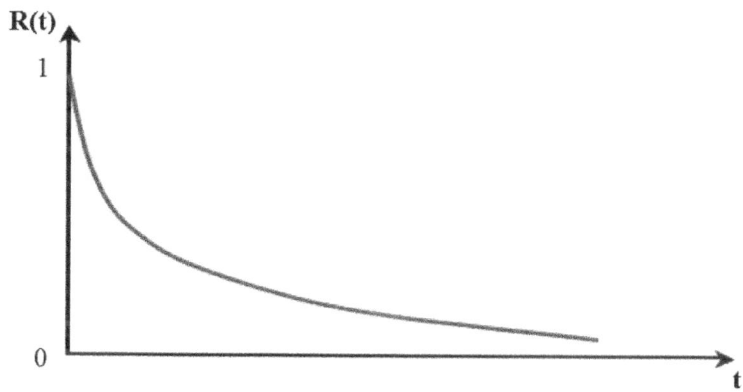

Figure 6.2. *Reliability function*

6.3. Failure in mechanics

For structures, the undesirable event is failure or non-compliance with the acceptability criteria, which generally corresponds to collapse or a limiting state. We write P_f for the probability of failure. The reliability is defined as: $1 - P_f$.

We will often use the reliability index β, which is defined as:

$$\beta = \Phi^{-1}(1 - P_f) = -\Phi^{-1}(P_f) \qquad [6.2]$$

where Φ^{-1} is the inverse function of the standard normal distribution.

6.3.1. *Failure scenarios*

In structural reliability, we will attempt to verify that the structure will meet the requirements for which it was designed throughout the duration of its lifetime. This verification is based on conditions characterizing proper operation that we define beforehand. These conditions can, for example, be derived from our knowledge of physics, mechanics, chemistry or biology and expert opinions.

We can only verify the reliability of a structure against a given failure scenario in a conditional sense. The limiting states or failure criteria used for design and dimensioning represent the state of our understanding of the structure we are studying and the actions that it experiences. A state is qualified as a failure state if the structure satisfies one of the failure criteria that define the mode of failure. Otherwise, it is said to be a properly functioning state.

Unknowns that can be modeled are represented by random variables, which are expressed in the form of a vector denoted as$X$. Realizations of this random vector will be denoted as x.

This random vector is associated with a joint probability density f_x, which also contains information about the correlations between the variables. Each failure criterion is associated with a function $G(X)$ of random variables, called the failure function or limiting state function. The function G is a deterministic function that takes random variables as inputs and yields a random variable. By convention, if x is a realization of X, $G(x) > 0$ represents a properly functioning state, $G(x) < 0$ represents a failure state and $G(x)$ is equal to 0 for any state belonging to the surface of limiting states.

The failure function can often be written in a marginal form, or as the difference between the strength R (or the system resources) and the value of the applied stress S:

$$G(X) = R(X_1, \ldots, X_i) - S(X_{i+1}, \ldots, X_n) \qquad [6.3]$$

Strength–stress methods lay the foundation for structural reliability methods. They are the simplest class of methods, and underpin dimensioning for all types of structure, whether mechanical, static, dynamic, etc.

The function G is not necessarily an explicit function of the initial variables X. It can depend on the structural response (for example stress and strains), in which case it requires prior mechanical calculations involving a constitutive law in order to be evaluated (which are usually done by finite elements).

6.3.2. *Expression of the failure probability*

A measure for the failure is then given by the probability of the event $\{R - S \leq 0\}$, i.e.:

$$P_f = \text{Prob}\,(R - S \leq 0)$$

Failure occurs if the strength is smaller than the stress.

6.4. Reliability index

The earliest proposal for a reliability index appears to be due to Rjanitzyne in the 1950s. After this point, various other versions were proposed, but the most complete was suggested by Hasofer and Lind, who established a rigorous definition. We will present the indices proposed by Rjanitzyne and Cornell and Hasofer and Lind.

6.4.1. *Rjanitzyne–Cornell index*

The β_c index is obtained from the mean m_z and the standard deviation σ_z of the marginal variable $Z = R - S$:

$$\beta_c = \frac{m_z}{\sigma_z} \qquad [6.4]$$

The β_c index therefore arises as the inverse of the variation coefficient of the random variable Z. β_c reflects the common engineering practice of introducing a shift of several standard deviations from the mean. The approach of the Rjanitzyne–Cornell index is only rigorous in the case where the variables are Gaussian and the limiting state is linear. Indeed, using a different representation for the marginal distribution Z but keeping the same limiting state leads to a different value for β_c.

This gives rise to two issues:

– the first is related to the non-invariance of the reliability index β_c for different representations of the same limiting state function. The index proposed by Hasofer and Lind resolves this issue;

– the second is related to the problem of deriving the failure probability from the index. This is achieved by using the first- and second-order approximations.

The β_c index can be viewed as an approximation of the exact reliability index. It is invariant under first-order expansion, but only if this expansion is performed around a specific point: the most likely point of failure. The β_c index is accurate in problems with Gaussian variables and a limiting state that is linear over physical space.

6.4.2. *Hasofer–Lind index*

To resolve the issue of the non-invariance of β_c, Hasofer and Lind suggested departing from a representation in the space of physical variables by performing a change in variables to a new space of statistically independent Gaussian variables with zero mean and unit standard deviation:

$$X_i \to U_i \quad \text{Gaussian vector} \quad N(0,1), \quad m_{U_i} = 0, \quad \sigma_{U_i} = 1, \quad \forall i, \forall j \qquad [6.5]$$

The transformation from the physical space to the normed space (or standard space) is immediate for independent Gaussian variables and conserves the linearity of the limiting state:

$$u = \frac{x - m_X}{\sigma_X} \quad \text{(reduced variables, which here are Gaussian)} \qquad [6.6]$$

6.5. Mechanoreliability coupling

Taking uncertainty into account in mechanical analysis is necessary if we wish for structural dimensioning to be both optimal and robust. This objective has driven the development of probabilistic approaches in structural mechanics over the last few decades. These methods allow us to study both the reliability of components and systems and the influence of parameter variability on the behavior of these components or systems [ELH 13a].

Mechanical models are often complex (nonlinear behavior, dynamics, fatigue, fracture mechanics, etc.) and require finite element codes in order to be solved. This complexity needs to be managed using mechanoreliability coupling methods before reliability analysis can be meaningful. This further increases the requirements in terms of computational resources. Indeed, coupling methods involve a call to a deterministic calculation of the mechanical model in order to draw useful values for the random variables. In general, the number of calls scales with the number of random variables, causing the computation time to become prohibitively large if traditional methods are used.

The primary objective is therefore to minimize the number of computations that must be performed while maintaining satisfactory levels of robustness and confidence in the results: the failure probability and its most significant influencing factors. In the following, we will present the basic principles of reliability-based computation, as well as the various different methods that are employed today. We will highlight the difficulties in implementing approaches that couple mechanics and statistics. We will see that different degrees of approximation are possible depending on the characterization of the limiting state, working from the concepts discussed above.

6.5.1. *Reliability-based calculation methods*

Mechanoreliability calculations incorporate the concept of probability measures. Implementing these calculations requires us to implement a certain mathematical formalism. This section describes the implementation of the limiting state function, random variables and the large classes of methods that will allow us to solve reliability problems.

Limiting state function

Let X be the vector of initial variables, which can be variables describing external actions, material characteristics or the geometric characteristics of a component or mechanical system. The risk associated with these variables is defined by their joint probability density. This random vector is defined in physical space. Performance functions or limiting state functions $G(X)$ are defined in such a way that $G(X) > 0$ is the safety domain, and $G(X) \leq 0$ is the failure domain for a given failure scenario. The boundary between the two domains, $G(X) = 0$, is called the limiting state.

To solve the reliability problem, there are two distinct classes of methods:

– The first class of methods is based on studying the performance function $G(X)$. The failure probability is defined as follows:

$$P_f = \text{Prob}(G \leq 0) \qquad [6.7]$$

To calculate this probability, we do not need to know the shape of the limiting state. This class includes simulation-based methods, most notably Monte Carlo (MC) simulations and Latin square simulations. Samples of the population of initial variables are generated to obtain a representative population, which serves as the basis for statistical analysis.

– Methods in the second class study the limiting state. This approach finds the shape of the boundary $G(X) = 0$. The goal is to estimate the following integral:

$$Pf = \int_{G \leq 0} f_X(x)dx \quad [6.8]$$

Studying the limiting state allows us to calculate the failure probability, but also lets us define other indicators such as reliability indices and also the design point known as the most likely point of failure. Finding this point provides additional information that is very useful in analyzing the structure and the most significant influencing factors associated with the design point. These indicators are not made available by methods that consider the performance function.

6.5.2. *Monte Carlo method*

MC methods use the law of large numbers to evaluate a deterministic object, typically an integral, that represents the mathematical expected value of a random vector $\{X\}$ with values in $\mathbb{R}^n$ based on N independent simulated realizations.

The values of the initial variables (the vector $\{X\}$) are randomly sampled using the probability distributions of $\{X\}$ (Figure 6.3). We count the number of draws N_f that fall within the failure domain D_f, i.e. the number of draws satisfying the condition $G(x) \leq 0$. The failure probability P_f is then evaluated as:

$$P_f = \int_{D_f} \Phi_n(u_k)du_1 \dots du_n \quad [6.9]$$

We can now introduce the failure indicator $I(.)$ defined by:

$$I(G(x)) = \begin{cases} 1 & \text{if} \quad G(x) \leq 0 \\ 0 & \text{if} \quad G(x) > 0 \end{cases} \quad [6.10]$$

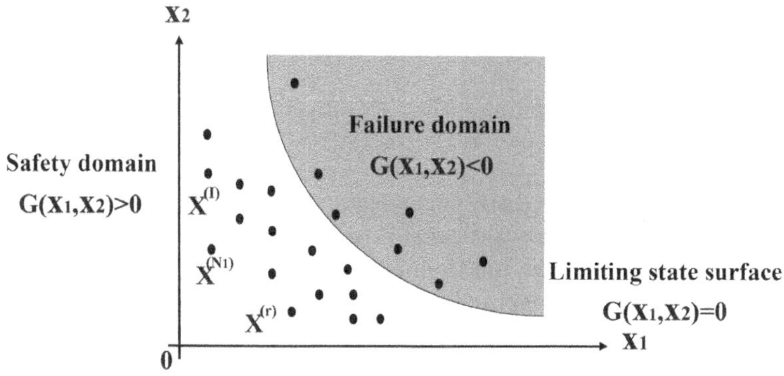

Figure 6.3. *Representation of the Monte Carlo method*

The probability of failure can now be given over the whole domain using the expression:

$$P_f = \int_D I(G(x))\Phi(u_k)du_1 \ldots du_n = E[I(G(x))] \quad [6.11]$$

We can now define an unbiased estimator of this expected value by taking the empirical average of $I(.)$ over N random draws:

$$P_f = E\left[I(G(x))\right] = \frac{1}{N}\sum_{i=1}^{N} I(G(x)) \quad [6.12]$$

The mean of this sample converges to the true mean P_f as the number of draws approaches infinity.

6.5.3. *FORM/SORM approximation methods*

The first-order reliability method (FORM) and second-order reliability method (SORM) are approximation methods that allow us to calculate the multidimensional integral [6.8]. These methods work in three steps:

– perform a transformation T of the variables from the physical space x_i to the Gaussian normed space u_i;

– determine the most likely point of failure P (design point);

– approximate the limiting surface by a hyperplane or a quadratic surface at the design point.

To allow us to more easily manipulate random variables, and since the Gaussian distribution possesses certain special properties, the first step of this method transforms (by the transformation T) the vector of random variables $\{x\}$ into a vector $\{u\}$ composed of normal random variables with zero mean and unit variance:

$$u_i = T(x_j) \quad \text{and} \quad H(u_i) = G(T^{-1}(u_j)) \qquad [6.13]$$

Figure 6.4 illustrates this transformation from the physical space to the normed space in the case of a random variable.

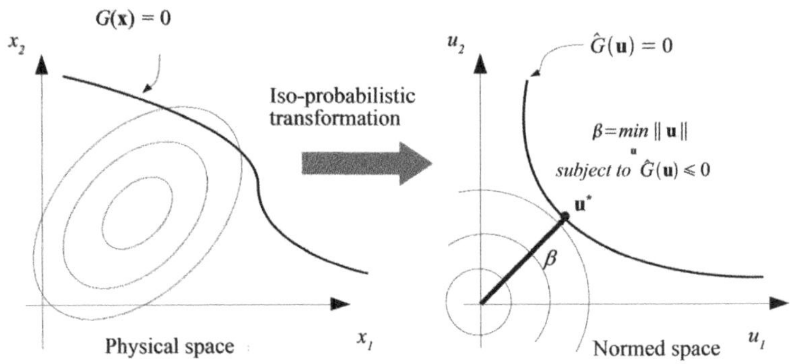

Figure 6.4. *Representation of the density curves and the limiting state function in the physical and normed spaces*

The choice of transformation is made based on the statistical information available on the random variables, the normality properties of the random variables and the dependencies that might exist between them.

The second step of the FORM/SORM methods is to determine the design point, i.e. the point in the failure domain that is closest to the origin in the normed space. In the normed space, the probability density contours are concentric spheres, and the probability density decreases with the distance from the origin. Therefore, the design point P^* is the failure point with the highest probability of occurring. The design point in the standard normal space, P^*, can be calculated in the physical space in order to give a more meaningful interpretation of this point. The design point is then the solution of the following nonlinear optimization problem:

$$\begin{cases} P^* = \min(\sqrt{\langle u, u \rangle}) \\ \text{satisfying} : H(u_k) \leq 0 \end{cases} \quad [6.14]$$

The third step of the FORM/SORM methods approximates the surface of limiting states in the space of normed variables by a tangent surface at the design point P^*, i.e. the closest point to the origin subject to the condition $H(u) = 0$. The distance between the origin and the design point, called the Hasofer–Lind reliability index and denoted by HL, represents a metric that describes the reliability of the system being considered (Figure 6.5). The reliability index may therefore be written as:

$$\beta_{HL} = u^* \alpha^* \quad [6.15]$$

where u^* and α^*, respectively, denote the coordinates and the vector formed by the direction cosines of the design point.

In the FORM, the surface of limiting states in the normed space is approximated by a tangent hyperplane at the design point. The failure probability may then be calculated from the equation:

$$P_f = \Phi(-\beta_{HL}) \quad [6.16]$$

This linear approximation of the surface of limiting states around the design point will only be accurate if the limiting state function is linear or weakly nonlinear.

For strongly nonlinear limiting state functions, FORM may therefore not be sufficient to find a reasonable value for the failure probability. In these cases, a better approximation of the surface of limiting states at the design point may be required. To do this, a second-order (parabolic) function is fitted to the nonlinear limiting state function at the design point.

This method is called the SORM. It is a complicated process and involves relatively high computation times. The method converts the limiting state function into the form of two independent random functions, one linear and the other quadratic in U-space. Because of our limited knowledge of the probability distribution of the quadratic term, an exact analytical calculation of the failure probability may not always be possible. An asymptotic approximation is used, and the failure probability is expressed as a function of the FORM reliability index (the HL reliability index β_{HL}):

$$P_{f_A}^{SORM} \simeq \Phi(-\beta_{HL}) \prod_{j=1}^{n-1} \frac{1}{\sqrt{1+\beta_{HL}.k_j}} \qquad [6.17]$$

In the above, the index A denotes the asymptotic approximation, n is the dimension of the normed space, and the k_j, where $j = 1$ to $(n-1)$, are the principal curvatures of the surface of limiting states at the design point. Breitung [BRE 89] showed that equation [6.17] asymptotically approaches the exact value of the failure probability as the values of β_{HL} tend to infinity if the product $(\beta_{HL}.k_j)$ is fixed. The asymptotic approximation, therefore, gives good results for large values of β_{HL}. However, the approximation accumulates large errors as the curvatures tend to $\frac{1}{\beta_{HL}}$, i.e. $\frac{1}{\beta_{HL}} \simeq k_j$, due to the singularity at this point.

6.6. Reliability-based optimization in mechanics

The objective of the design phase, as mentioned earlier, is to establish a definition of best performance that achieves a reasonable compromise between contradictory needs, such as reliability and cost. It is therefore necessary to integrate reliability-based optimization procedures into the design of dynamic structures by using numerical modeling software that allows us to analyze the numerous possibilities and that automates the search for the optimal form, and by integrating reliability analysis into the process to evaluate the performance of the optimal design [ELH 13b].

Several methods have been used in the literature to solve the RBDO problem in dynamics, the most notable of which is the hybrid method [MOH 06]. This method was developed by Kharmanda [KHA 02] by merging both the physical and normal spaces. Studies have been conducted on structures with different parameters (stresses, strains and frequencies). The next objective is to give an integrated account of the phenomenon of resonance in mechanical structures.

The hybrid method has been used for vibrational dynamics computations in applications with the aim of finding the critical band around the resonance frequency. In these applications, executing the method was highly complex and incurred very high computation times. To overcome these two obstacles, another method known as the safest point (SP) method has been developed, which effectively finds the optimal reliable solution for vibrating structures.

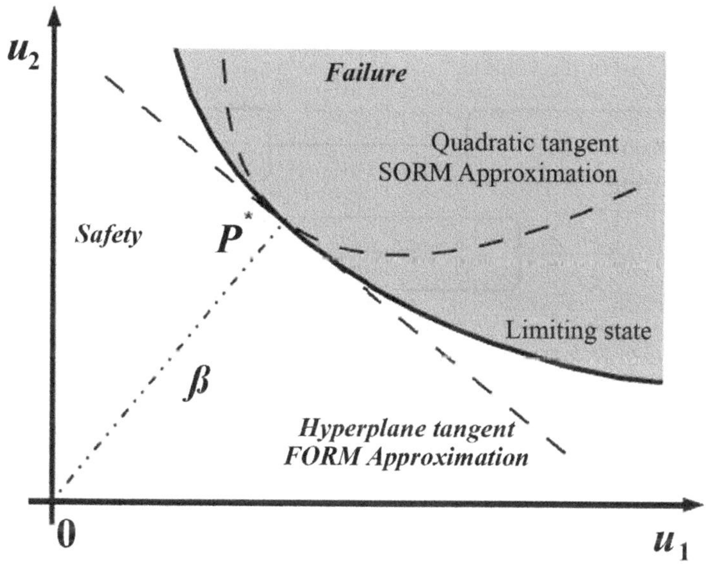

Figure 6.5. *Approximation with the FORM and SORM methods*

In the following, we will summarize the various different reliability-based optimization methods for designing vibrating structures, and evaluate these methods in the example of a numerical application to fluid–structure interactions in order to demonstrate the effectiveness and robustness of both the presented methods and the proposed method.

6.6.1. *Deterministic optimization*

In deterministic optimization, the designer introduces safety coefficients with the objective of improving the safety of the design. These coefficients are used by the most critical structural parameters, such as the load. These coefficients are chosen based on experiments or representative testing or are calibrated based on probabilistic schemes.

The deterministic design optimization process is illustrated in Figure 6.6. The deterministic optimization loop involves three steps:

1) describe the geometry by a geometric model (using CAD software);

2) analyze the model using the finite element method;

3) optimize the shape of the structure by minimizing an objective function such as the cost or the volume.

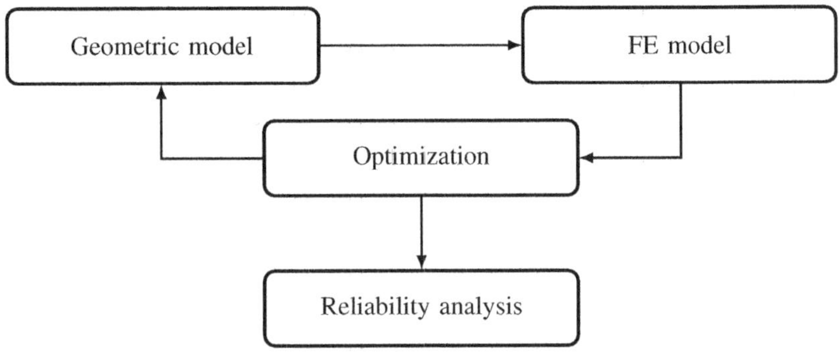

Figure 6.6. *Deterministic design optimization process*

Deterministic structural optimization requires a high number of instances of structural analysis and verifications of failure modes. But in reality, design obstacles are encountered when attempting to describe the uncertainty of the variables, since the optimal solutions can decrease the reliability level. This justifies the need for reliability-based optimization.

6.6.2. *Different approaches to RBDO*

The safety coefficients are calibrated over a large class of structures, and their application in the optimization does not guarantee that the optimum will have adequate reliability. In fact, the reliability of the obtained optimum is entirely neglected, as these coefficients are not directly related to the reliability requirements. The result of this is that the optimal solution may lack robustness. Reliability-based optimization analyzes the failure at each iteration in order to determine the role of uncertainty in the system. However, in practical applications, the coupling between the reliability analysis and the optimization procedure leads to very high computation times and low stability in the convergence. In the space of random variables, solving the reliability problem generates large numbers of calls to the mechanical model, whereas in the space of design variables finding the optimal solution modifies the configuration of the structure and therefore requires the reliability to be re-evaluated at each iteration. There is therefore a strong need to develop effective techniques for reducing the computation times.

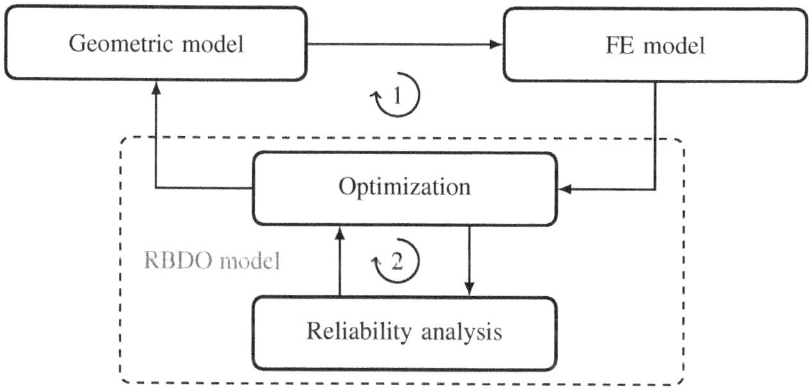

Figure 6.7. *Sequential process of reliability-based design optimization*

In the RBDO process (Figure 6.7), there are two nested loops: an initial optimization loop for an objective function such as the cost and the volume subject to certain conditions, e.g. physical, geometric, functional, etc., and a second reliability computation loop involving an optimization procedure for a given limiting state.

To introduce reliability into the design optimization, we view the reliability index as a constraint that must be satisfied. In the classical deterministic optimization process, safety coefficients are introduced for the most significant variables (load, geometry, etc.). Any optimization algorithm can then be used to find the optimal solution. At the optimal point, reliability analysis is performed in order to determine the reliability index of the limiting state that is being considered. In the classical reliability-based optimization process, safety coefficients are replaced by reliability constraints, which are directly integrated into the main optimization problem. This requires the reliability index and the most likely point of failure to be re-evaluated at each iteration in order to ensure that the target reliability threshold is observed.

Reliability-based optimization represents a new approach that introduces reliability criteria into the search for the optimal configuration. Its objective is to establish a definition of the best compromise between reducing the cost and guaranteeing the reliability while integrating an account of the uncertainty associated with the system variables. The optimization procedure needs to allow the role of the uncertainty to be redistributed based on the sensitivity of the design objectives. Because of this, deterministic optimization is not capable of providing suitable levels of reliability. Deterministic design can lead to high economic losses as a result of failure, whereas excessively reliable design can often generate excessive costs. RBDO aims to find a balanced design while reducing the overall cost.

This approach has the advantage of beginning the search for the optimum by adjusting the variables, which are penalized independently of their mechanical roles. In this sense, system robustness is achieved, since the role of uncertainty is reduced over the course of the optimization procedure.

From a numerical perspective, reliability-based optimization approaches require a coupling between the mechanical model (finite element method), the optimization algorithms and the structural reliability analysis tools. Classically, reliability analysis is integrated into optimization methods in two spaces: the normed space of random variables and the physical space of design variables, which incurs an extremely high computational cost.

In the reliability-based formulation, we distinguish between two types of variable:

1) the optimization variables x, which are deterministic variables that must be adjusted in order to optimize the dimensioning; they represent the control parameters of the mechanical system (i.e. dimensions, materials, loads, etc.) and the probabilistic model (i.e. mean values and standard deviations of random variables);

2) the random variables y, which represent the uncertainty associated with the design variables. Each of these variables is identified by its distribution law and corresponding parameters. These variables can be geometric dimensions, material characteristics, or external loads.

6.6.3. *Classical approach*

Classical RBDO performs the reliability analysis first. The solution thus obtained is used to minimize the objective function subject to physical, geometric or functional constraints, including the reliability constraint. Convergence is attained after a certain number of iterations, alternating between the optimization and the reliability procedures. These problems can be solved using any nonlinear programming algorithm, such as the sequential quadratic programming (SQP) method or the penalty method.

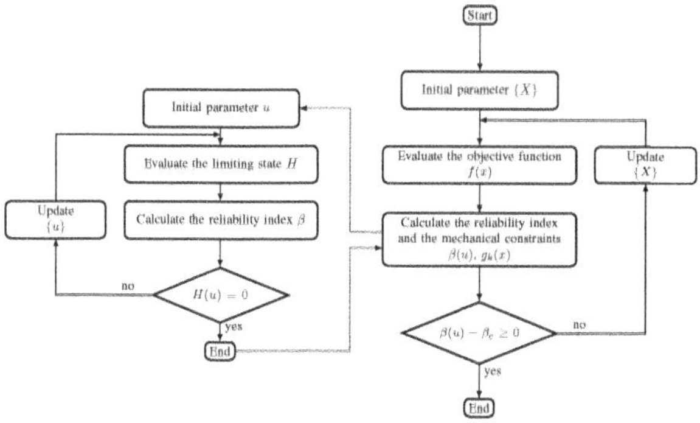

Figure 6.8. *Algorithm of the classical approach*

In this approach, the total number of iterations is found by summing the iterations of both problems: optimization and reliability, which leads to a very large number of evaluations of the mechanical model [FEN 86, FRA 94]. To overcome this difficulty, we introduce the hybrid formulation of reliability-based optimization.

6.6.4. *Hybrid approach*

This approach integrates the two subproblems into a single problem that is solved simultaneously in the spaces of both deterministic and random variables. This scheme integrates the reliability problem into a formulation of the optimization problem in order to achieve a single objective function $F(x, y)$ and therefore to reduce the overall computational cost. It can be written in the form:

$$F(x, y) = f(x) \cdot d_\beta(x, y) \qquad [6.18]$$

where $f(x)$ and $d_\beta(x, y)$, respectively, denote the objective function and the image of the reliability index in the hybrid space (Figure 6.9).

All information associated with the optimization process is gathered together in this space; P_y^* is the optimal solution of the problem and P_x^* is the design point.

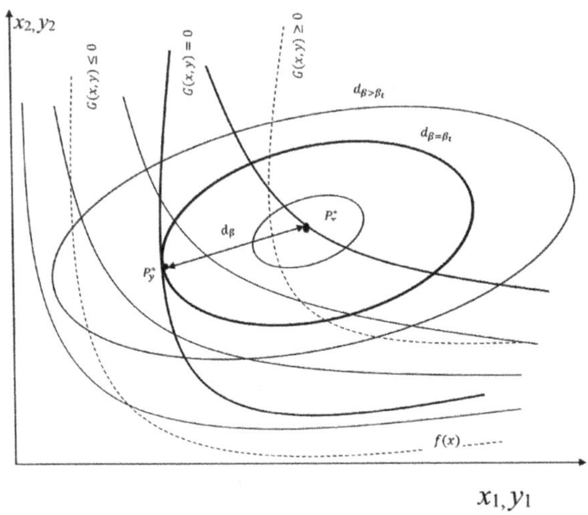

Figure 6.9. *Hybrid design space*

The hybrid RBDO problem has the following expression:

$$\begin{cases} \min\limits_{x,y} : F(x,y) \\ \text{s.t.} : G(x,y) \leq 0 \\ \quad g_k(x) \leq 0 \\ \quad d_\beta(x,y) \geq \beta_c \end{cases} \qquad [6.19]$$

where $G(x,y) \leq 0$ is the limiting state, $g_k(x) \leq 0$ are the deterministic constraints and β_c is the target reliability index.

The objective of the new problem is to minimize the function $F(x,y)$ subject to the active constraint $G(x,y) \leq 0$, which must strictly be met, and the other constraints $g_k(x) \leq 0$ and $d_\beta(x,y) \geq \beta_c$, which are the available resources to be used as needed. In traditional approaches, the reliability index is calculated using an optimization procedure in the standard Gaussian space. The hybrid formulation, on the other hand, considers both spaces:

deterministic and random. The solution is calculated in the hybrid design space (x, y). This procedure leads the two problems to converge in parallel.

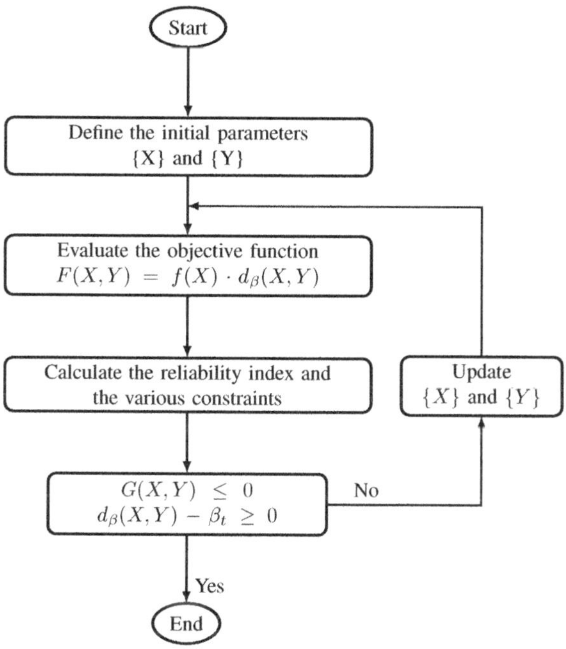

Figure 6.10. *Algorithm of the hybrid method*

6.6.5. *Frequency-based hybrid approach*

Integrating reliability-based optimization procedures leads to a new method called the adapted hybrid dynamic method for freely vibrating structural problems [MAK 08]. The main idea is to look for two design points instead of a single point.

These two points are used to find lower and upper bounds for the critical frequency band. This zone coincides with the crucial frequency band in the neighborhood of the resonance frequency of the structure being optimized (Figure 6.11). These two bounds are identified at each iteration of the computation.

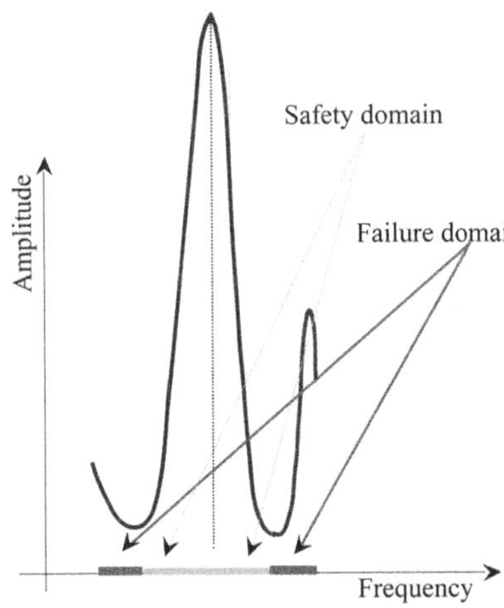

Figure 6.11. *Resonance frequency. For a color version of this figure, see www.iste.co.uk/elhami/interactions.zip*

Formulation of the optimization problem

The formulation of the dynamic vibration problem is given by:

$$\begin{cases} \min\limits_{x,y} : F(x,y) = f(x) \cdot d_{\beta_A}(x,y) \cdot d_{\beta_B}(x,y) \\ \quad \text{s.t.} G(x,y) \leq 0 \\ \qquad g_k(x) \leq 0 \\ \qquad d_{\beta_A}(x,y) \geq \beta_t \quad \text{and} \qquad d_{\beta_B}(x,y) \geq \beta_t \end{cases} \qquad [6.20]$$

$d_{\beta_A}(x,y)$ is the image of the reliability index in the hybrid design space (HDS), $G(x,y) \leq 0$ is the limiting state, $g_k(x)$ are the constraints and β_t is the target reliability index.

Case of a normal distribution

For normal distributions, the problem has two design variables, which are random variables. In the normal space, the distance $d(u)$ is defined by $d(u) = \sqrt{\Sigma u_i^2}$, where u_i is a normal variable defined by:

$$u_i = \frac{y_i - m_i}{\sigma_i}, \qquad i = 1, \ldots, n \qquad [6.21]$$

We replace the mean m_i by the deterministic design variable x_i ($x_i = m_i$). The image of the reliability index in the hybrid space is defined by:

$$d_\beta = \sqrt{\sum_{i=1}^{n} \left(\frac{y_i - x_i}{\sigma_i}\right)^2} \qquad i = 1, \ldots, n \qquad [6.22]$$

The reliability levels d_β are represented by ellipses (in the case of a normal distribution).

Case of a log-normal distribution

In many mechanical problems, random variables cannot take negative values due to physical aspects of the problem. In this situation, the log-normal distribution is more appropriate, since it automatically eliminates the possibility of negative values. This distribution has direct links with the normal distribution. Indeed, if the random variable X has a log-normal distribution, then $\ln(X)$ has a normal distribution characterized by two distribution parameters, the mean m_i and the standard deviation σ_i. The normal variable u_i is defined by:

$$u_i = \frac{\ln(y_i) - \mu_i}{\xi_i}, \quad i = 1, ...n \qquad [6.23]$$

where μ_i and ξ_i are the two distribution parameters of the log-normal distribution. They are defined by:

$$\mu_i = \ln\left(\frac{x_i}{\sqrt{1 + \gamma_i^2}}\right), \qquad i = 1, \ldots, n \qquad [6.24]$$

$$\zeta_i = \sqrt{\ln(1 + \gamma_i^2)}, \qquad i = 1, \ldots, n \qquad [6.25]$$

where $\gamma_i = \frac{\upsilon_i}{x_i}$.

The normal variable u_i can also be expressed as:

$$u_i = \frac{\ln\left(\frac{y_i\sqrt{1+(\sigma/x_i)^2}}{r_i}\right)}{\sqrt{\ln(1 + (\sigma/x_i)^2)}}, \quad i = 1, \ldots, n \qquad [6.26]$$

and the image of the reliability index in the hybrid space is defined by:

$$d_\beta = \sqrt{\sum_{i=1}^{n} \frac{\left(\ln\left(\frac{y_i\sqrt{1+(\sigma/x_i)^2}}{x_i}\right)\right)^2}{\ln(1+(\sigma/x_i)^2)}}, \qquad i = 1, \ldots, n \qquad [6.27]$$

6.7. SP method

The approach presented in the following section was developed for applications with dynamic vibration computations with the objective of finding the critical frequency bands for structural optimization problems. It is known as the SP method [ELH 15]. The underlying idea of this proposed new technique is to find an optimal and reliable solution for structures that are both vibrating and interacting with a flowing fluid. The desired solution is given by a point with the same reliability index relative to the lower and upper bounds of the critical frequency band $[f_a, f_b]$.

6.7.1. *Formulation of the problem*

Consider the equality obtained by equating both reliability indices: $\beta_a = \beta_b$ with

$$\beta_a = \sqrt{\sum_{i=1}^{n} (u_i^a)^2} \quad \text{and} \quad \beta_b = \sqrt{\sum_{i=1}^{n} \left(u_i^b\right)^2} \qquad i = 1, \ldots, n \qquad [6.28]$$

To satisfy equation [6.28], we establish the equality in each limit. Thus, we have:

$$u_i^a = -u_i^b \qquad i = 1, \ldots, n \qquad [6.29]$$

We will perform two separate studies for the normal distribution and the log-normal distribution.

Case of a normal distribution

For a normal distribution, the normalized variable u_i has the following form:

$$u_i = \frac{y_i - m_{y_i}}{\sigma_i}, \qquad i = 1, \ldots, n \qquad [6.30]$$

where the mean m_{y_i} and the standard deviation σ_i are the two parameters of the normal distribution.

From equation [6.29], we have that:

$$\frac{y_i^a - m_i}{\sigma_i} = -\frac{y_i^b - m_i}{\sigma_i}, \text{ or } \frac{y_i^a - x_i}{\sigma_i} = -\frac{y_i^b - x_i}{\sigma_i}, \quad \mathrm{i} = 1, ..., n [6.31]$$

The SP, which satisfies equality in the two reliability indices ($\beta_a = \beta_b$), is located in the center of the interval $[f_a, f_b]$ and corresponds to the frequency f_n:

$$m_i = x_i = \frac{y_i^a + y_i^b}{2}, \quad \mathrm{i} = 1, \ldots, \mathrm{n} \qquad [6.32]$$

Case of a log-normal distribution

For a log-normal distribution, the normal variable u_i is defined by:

$$u_i = \frac{\ln(y_i) - \mu_i}{\xi_i}, \quad i = 1, \ldots, n \qquad [6.33]$$

where μ_i and ξ_i are the two distribution parameters of the log-normal distribution, which are defined by:

$$\mu_i = \ln\left(\frac{x_i}{\sqrt{1+\gamma_i^2}}\right), \qquad i = 1, \ldots, n \qquad [6.34]$$

$$\zeta_i = \sqrt{\ln(1+\gamma_i^2)}, \qquad i = 1, \ldots, n \qquad [6.35]$$

where $\gamma_i = \frac{\sigma_i}{x_i}$.

The normal variable u_i can be expressed as:

$$u_i = \frac{\ln\left(\frac{y_i\sqrt{1+(\sigma/x_i)^2}}{x_i}\right)}{\sqrt{\ln(1+(\sigma/x_i)^2)}}, \quad i = 1, \ldots, n \qquad [6.36]$$

From equation [6.29], we have that:

$$\frac{\ln(y_i^a) - \mu_i}{\xi_i} = -\frac{\ln(y_i^b) - \mu_i}{\zeta_i}, \qquad i = 1, \ldots, n \qquad [6.37]$$

where $\mu_i = \ln\left(\frac{x_i}{\sqrt{1+\gamma_i^2}}\right) = \frac{\ln(y_i^a)+\ln(y_i^b)}{2}$.

In this case, the SP, corresponding to the frequency f_n and located in the interval $[f_a, f_b]$, is given by:

$$m_i = x_i = \sqrt{1+\gamma_i^2}\exp\left(\frac{\ln(y_i^a \cdot y_i^b)}{2}\right), \qquad i = 1, \ldots, n \qquad [6.38]$$

Implementing the SP method

The SP method can be expressed in three steps:

Step 1: determine the first design point (or the most likely point of failure): the optimization problem in physical space minimizes the objective function of the first model subject to the constraint associated with the frequency f_a (f_a is the lower bound of the critical frequency band). The resulting solution is considered to be the most likely point of failure and is taken as design point A.

Step 2: determine the second design point: the second optimization problem minimizes the objective function of the second model subject to the constraint associated with the frequency f_b (f_b is the upper bound of the critical frequency band). The resulting solution is considered to be the most likely point of failure and is taken as design point B.

Step 3: calculate the optimal solution: in this final step, we find the optimal design of the structure by means of a simple computation. This solution is given by the point corresponding to the resonance frequency f_n, which has the same reliability index relative to the lower and upper bounds of the critical frequency band $[f_a, f_b]$.

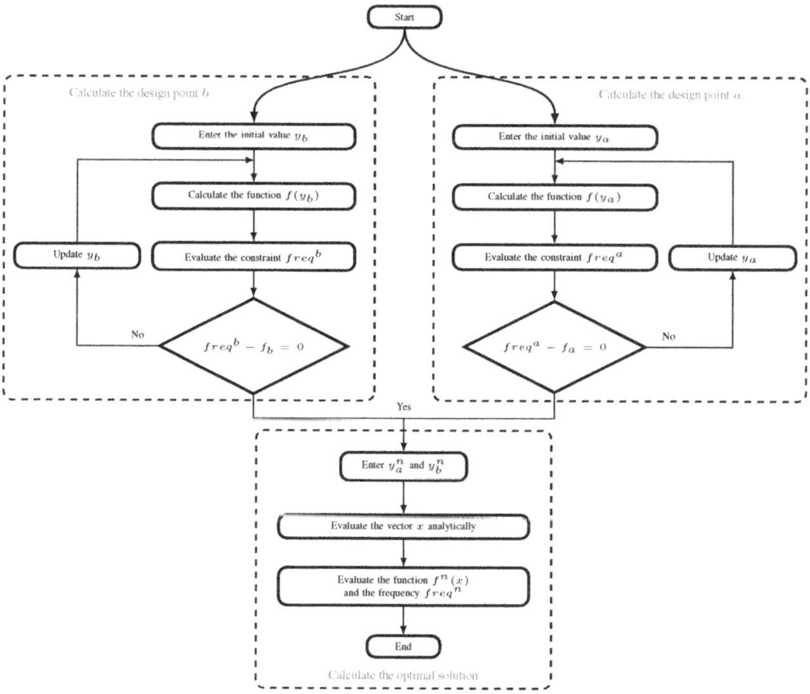

Figure 6.12. *Implementation of the SP method*

6.8. Numerical results

6.8.1. *Reliability calculation for an airplane wing*

Probabilistic study

In this application, we will apply the above results to the example of an airplane wing interacting with a flowing fluid in a problem modeled by finite elements/finite elements. Naturally, the results of deterministic analysis are only as good as the hypotheses and input data used in the analysis. The validity of the results depend on the values taken by the component in real conditions. In reality, every model analysis is subject to some uncertainty [ELM 15].

It is neither physically nor financially possible to fully eliminate the spread in the input parameters. Reducing the spread generally incurs higher costs, arising from either better-performing or more accurate manufacturing methods and improved quality control. Probabilistic studies are used to determine the effect of one or more variables on the analysis results. Given the complexity of the problem, and for reasons of simplification, we chose to only consider sources of uncertainty relating to the other components of the structure (geometry, boundary conditions, etc.).

Table 6.1 contains the mean values of the random variables and their standard deviations used for this study, as well as their chosen distributions.

Parameter	Mean	Standard deviation	Distribution
Young's modulus (Pa)	$7.1e10$	$0.355e10$	Gaussian (μ, σ)
Density (kg.m^{-3})	$2,770$	83.1	Gaussian (μ, σ)

Table 6.1. *Random variables and distributions*

Here, the stochastic calculations were performed using the probabilistic design system in Ansys©. This tool performs calculations using MC simulation for 100 samples and response surface methodology (RSM) for nine samples.

Table 6.2 shows the means and standard deviations of the natural frequencies of the wing with the airflow.

Modes	Deterministic case	MC	SD	RSM	SD
F_1	12.630	12.540	0.234	12.540	0.421
F_2	19.182	19.023	0.301	19.022	0.548
F_3	51.426	50.812	1.103	50.821	1.975
F_4	72.938	72.290	1.089	72.312	2.003
F_5	89.837	88.882	1.815	88.900	3.253
F_6	126.02	125.351	2.805	124.28	5.011

Table 6.2. *Natural frequencies of the wing*

Reliability analysis

In this numerical study, the proposed reliability-based methods used to calculate the failure probability are the FORM and SORM methods, which consider the uncertainty in the structural properties. Numerical computations are performed using a MATLAB© coupling code to evaluate the reliability of the structure, and using Ansys to calculate the implicit limiting state function G, which is defined as a function of the first natural frequency F_1 of the coupled system and limited by F_0 such that: $G(E, \rho) = F_1 - F_0$ with $F_0 = 11.8$ Hz.

Table 6.3 presents the design parameters and their statistical moments in the coupled wing, as well as the results obtained by the FORM and SORM methods.

Parameters	FORM	SORM
Young's modulus (Pa)	7.793e10	7.7927e10
Density (Kg/m^3)	3007.1	3007.2
Reliability index β	3.4575	3.5126
P_f	0.00027258	0.0002219
Reliability	99.973	99.978

Table 6.3. *Random variables and their statistical moments*

6.8.2. *Application of RBDO to the airplane wing*

The ONERA M6 wing was designed by Bernard Monnerie and his colleagues in 1972 as part of a collaboration within the Advisory Group for Aerospace Research and Development (AGARD) to serve as an experimental medium for three-dimensional flow studies at transsonic speeds and with high Reynolds numbers, and to validate numerical simulations of flows (representative conditions for real civilian and military flights). It has a symmetric profile, and its dimensions are listed in Figure 6.14.

Figure 6.13. *The M6 wing in the S2MA air tunnel at ONERA*

The results of these tests were published in 1979 in an AGARD report by Volker Schmitt and François Charpin [SCH 74]. The database with the results of the ONERA-M6 wing has been used hundreds of times to validate CFD software and is still widely used around the world. It is one of the most popular test cases, and is particularly well suited to understanding and evaluating modes of laminar-turbulent transitions, shock wave–boundary layer interactions, peeling, etc., which are characteristic of the phenomena that occur on wings near the speed of sound.

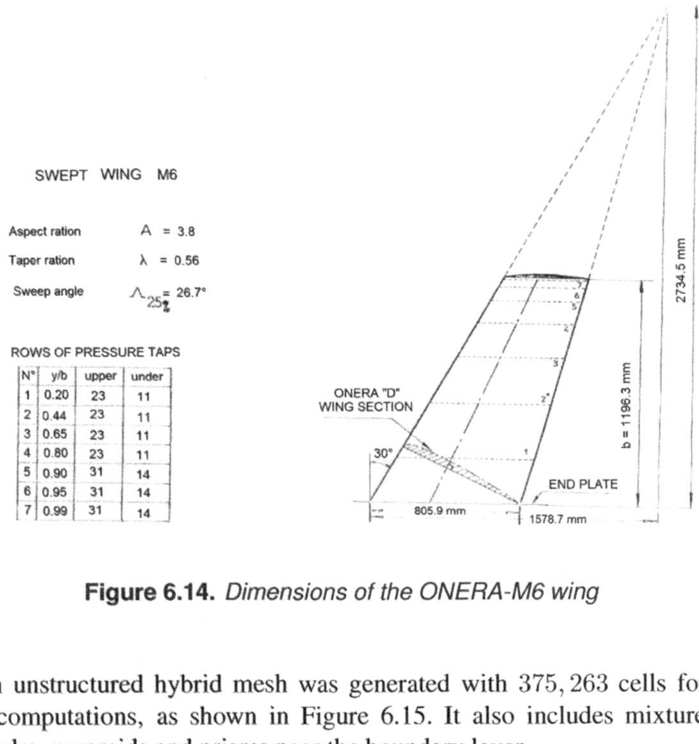

Figure 6.14. *Dimensions of the ONERA-M6 wing*

An unstructured hybrid mesh was generated with $375,263$ cells for the CFD computations, as shown in Figure 6.15. It also includes mixtures of tetrahedra, pyramids and prisms near the boundary layer.

Aerodynamic analysis of the ONERA M6 wing

Aerodynamic analysis was performed with the Ansys/Fluent solver with Mach number (M) 0.8395, angle of attack (α) $3.06°$ and Reynolds number (Re) $11.72E6$.

The pressure distribution on the upper surface of the wing is shown in Figure 6.16. A violent shock is observed in the leading edge near the fixed end, which diminishes toward the free end of the wing. Comparing the drag coefficient and the lift coefficient (see Figure 6.17) with those obtained by the NASA simulation calculated using the WIND code [SLA 02], we observe good agreement in the results for the refined mesh.

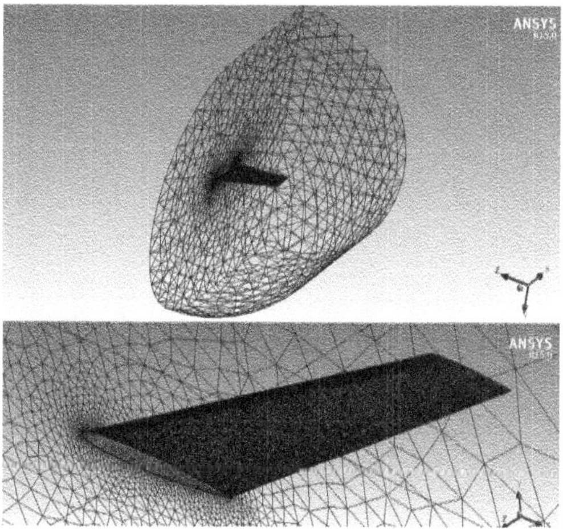

Figure 6.15. *Fluid mesh. For a color version of this figure, see www.iste.co.uk/elhami/interactions.zip*

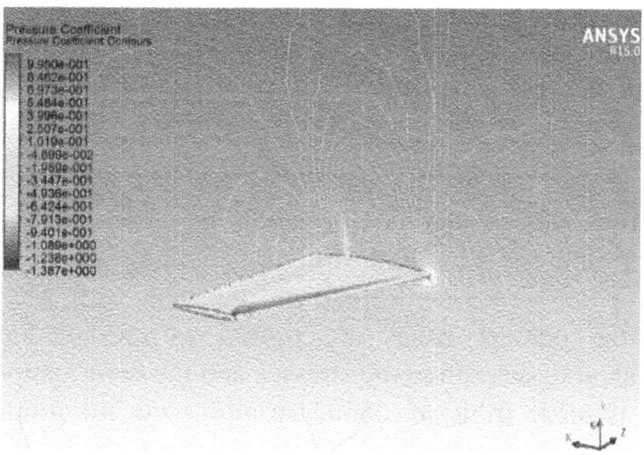

Figure 6.16. *Pressure distribution on the wing. For a color version of this figure, see www.iste.co.uk/elhami/interactions.zip*

	NASA CFD	Original mesh	Refined mesh
Cl	0.1410	0.1279	0.134
Cd	0.0088	0.0111	0.0096

Table 6.4. *Comparison of the drag coefficient (Cd) and lift coefficient (Cl)*

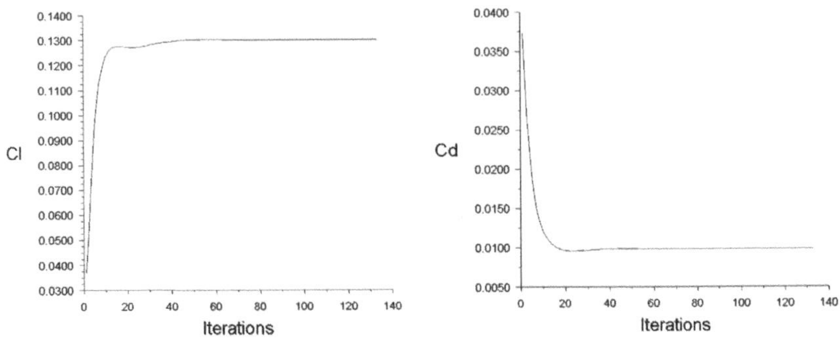

Figure 6.17. *Evolution of the drag and lift coefficients of the ONERA M6 wing as a function of the number of iterations*

We validate the simulation results by comparing the results found by Ansys/Fluent with the experimental data of the Onera M6 wing. We plot different contours of the pressure coefficient on cross-sectional planes of the wing, and compare the results of the simulation and the experimental data at positions $\frac{y}{b} = 0.2, 0.44, 0.65, 0.9$ and 0.95 [SCH 74], first for a coarse mesh, and then for our more refined mesh. We observe good agreement in the obtained results (see Figure 6.18).

Aeroelastic analysis of the Onera M6 wing

In this section, we consider a deformation due to aerodynamic load on the wing by performing one-way fluid–structure coupling with a static structure. After modeling the aerodynamic load on the wing with Ansys/Fluent, the pressures at the contact zones are passed as pressure-based loads to Ansys/Mechanical to determine the stresses and the deformation of the wing. The structure is made from an aluminum alloy with the following material properties: Young's modulus (E) 71 GPa, Poisson coefficient (ν) 0.32, and material density $2,770$ kg/m^3.

The calculated displacements of the wing in its deformed and non-deformed states are shown in Figure 6.19. We can clearly see that the calculated fluid pressure was successfully transferred and induced a deformation of the CSD volume mesh.

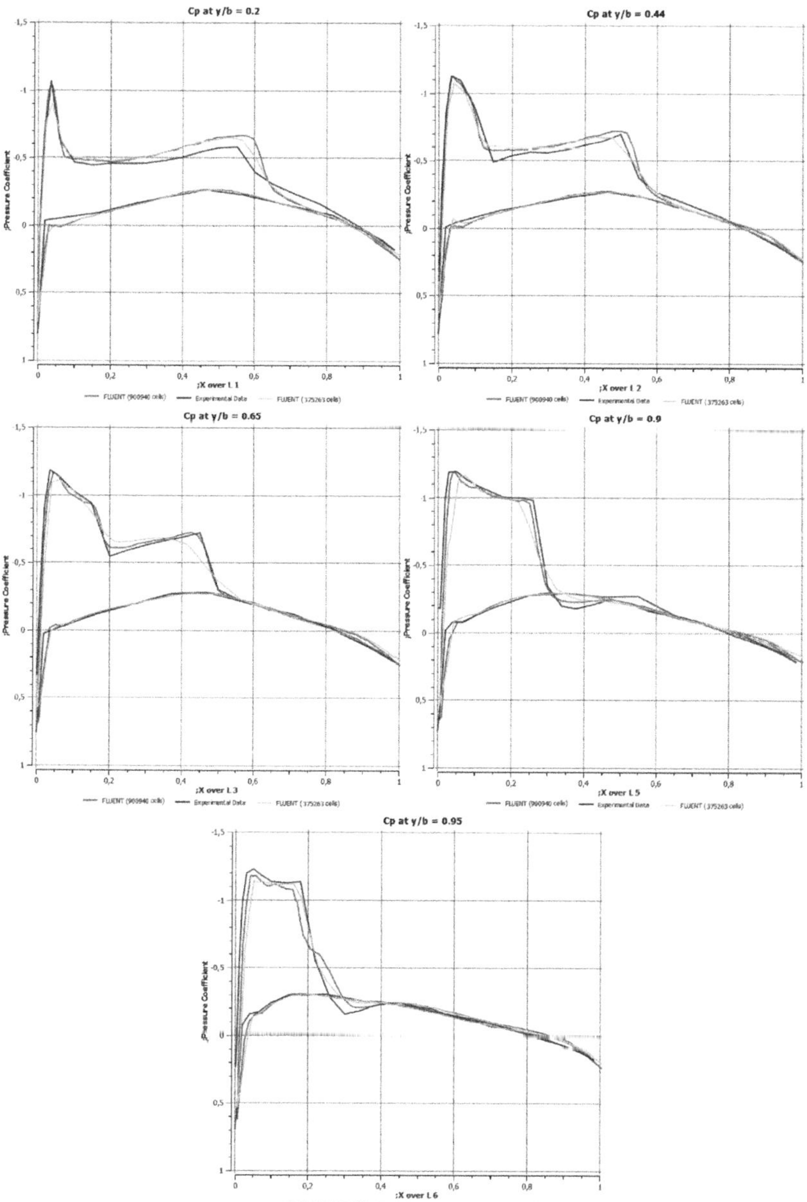

Figure 6.18. *Distribution of the pressure coefficient on the cross-sectional planes contours of the M6 wing as a function of the x-coordinate at the intrados and at the extrados. For a color version of this figure, see www.iste.co.uk/elhami/interactions.zip*

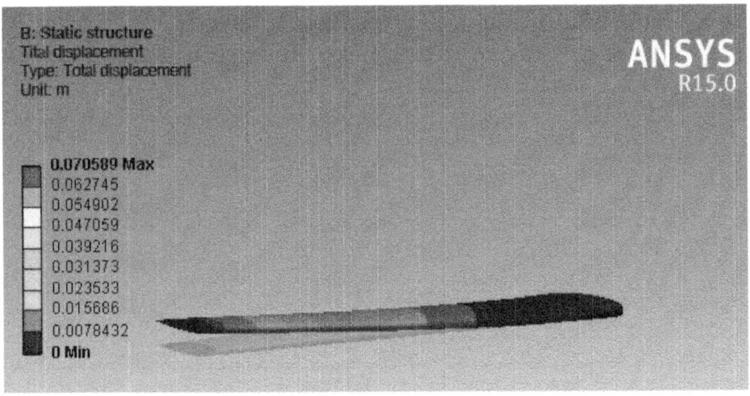

Figure 6.19. *Total displacement of the wing. For a color version of this figure, see www.iste.co.uk/elhami/interactions.zip*

Prestressed modal analysis

The eigenvalue and eigenvector problem is solved from the following equation:

$$[K]\phi_i = \lambda_i[M]\phi_i \qquad [6.39]$$

where $[K]$ is the stiffness matrix of the structure, ϕ_i is the eigenvector, λ is the eigenvalue and $[M]$ is the mass matrix of the structure. In prestressed modal analysis, the matrix $[K]$ incorporates the stress stiffness matrix. The results of modal analysis are shown in Figure 6.20. The results include the first four mode shapes with their natural frequency values.

Reliability-based design optimization

In the following, a comparative study of the hybrid RBDO method and the new SP method is given based on this application. We examine the advantages of the SP method using two probability distributions, the normal and log-normal distributions, to minimize the cross-section of the airplane wing subject to a frequency constraint. Other studies have been conducted in [ELM 15].

The dimensions of the cross-section are A, B and D, as shown in Figure 6.21 and Table 6.5.

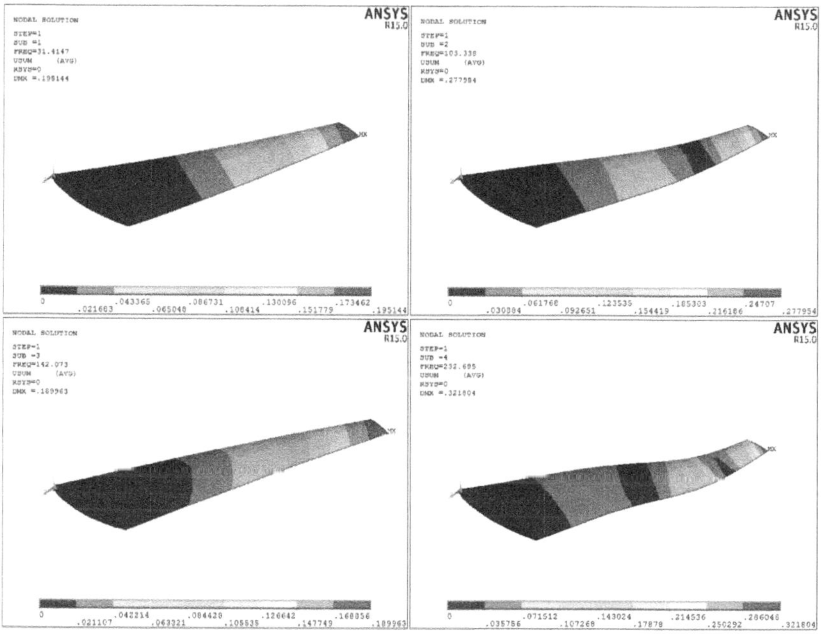

Figure 6.20. *Natural modes of the wing. For a color version of this figure, see www.iste.co.uk/elhami/interactions.zip*

Variables	A_N	B_N	C_N	D_N
Dimensions (m)	0.054	0.092	0.096	0.044

Table 6.5. *Parametrization of the wing*

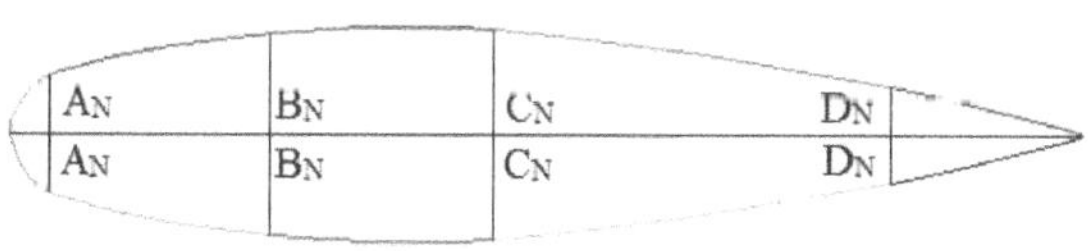

Figure 6.21. *Dimensions of the cross-section*

The primary objective is, for a given interval $[25, 35]$, to find the natural frequency located at the safest position in this interval. Hence, $F_a = 25$ Hz, $F_B = 35$ Hz and $F_n = ?Hz$, where f_n has to satisfy the reliability indices equation: $\beta_A = \beta_b$. We can now calculate the three models: the first structure must be optimized subject to the constraint of the first given value for the frequency F_a, the second must be optimized subject to the constraint of the second frequency F_B and the third structure must be optimized subject to the constraint of the frequency F_n that satisfies the reliability indices equation relative to both endpoints of the stated interval (Figure 6.22). Table 6.6 summarizes the parameters used for the three models.

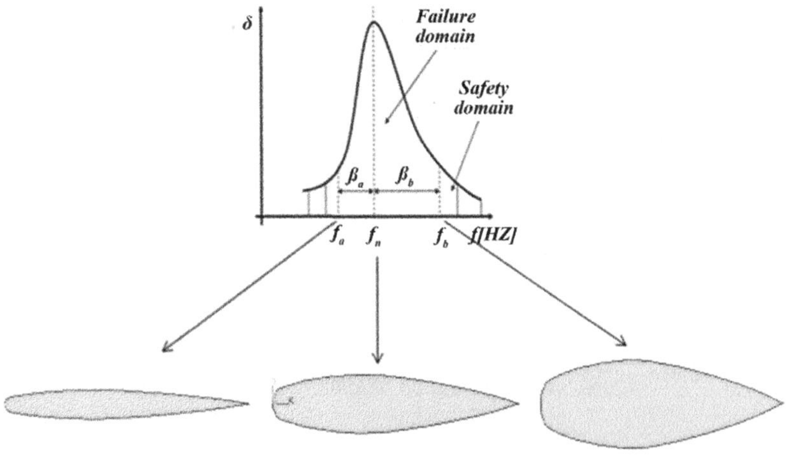

Figure 6.22. *Cross-sections of the three models. For a color version of this figure, see www.iste.co.uk/elhami/interactions.zip*

Hybrid method

We minimize the volume of the aircraft wing subject to the constraint of the first natural frequency and the reliability constraint. Formulating the problem with the hybrid method gives:

$$\begin{cases} \min\limits_{A_1,\ldots,m_A,\ldots,A_2,\ldots} & : \text{Volume}(m_A, m_B, m_C, m_D) \cdot d_{\beta_1}(A_1, B_1, C_1, D_1, m_A, m_B, m_C, m_D) \cdot d_{\beta_2}(A_2, B_2, C_2, D_2, m_A, m_B, m_C, m_D) \\ \text{s.t.} & : f_a(A_1, B_1, C_1, D_1) - f_a = 0 \quad , \quad \mathrm{f_b}(\mathrm{A_2}, \mathrm{B_2}, \mathrm{C_2}, \mathrm{D_2}) - \mathrm{f_b} = 0 \\ & : d_{\beta_1}(A_1, B_1, C_1, D_1, m_A, m_B, m_C, m_D) - d_{\beta_2}(A_2, B_2, C_2, D_2, m_A, m_B, m_C, m_D) \leq 0 \\ & : u_i^a + u_i^b = 0, \quad i = 1, \ldots, 4 \end{cases} \qquad [6.40]$$

where A, B, C and D are combined into the random vector $\{y\}$, with means given by the nominal values m_A, m_B, m_C, and m_D combined into a vector denoted $\{x\}$, with standard deviation 0.1.

<table>
<tr><th colspan="2">Variables</th><th>Parameters</th><th>Minimum value (MIN)</th><th>Maxmium value (MAX)</th></tr>
<tr><td rowspan="12">Design variables</td><td rowspan="4">F_N</td><td>A_N</td><td>0.027</td><td>0.081</td></tr>
<tr><td>B_N</td><td>0.046</td><td>0.138</td></tr>
<tr><td>C_N</td><td>0.048</td><td>0.144</td></tr>
<tr><td>D_N</td><td>0.022</td><td>0.066</td></tr>
<tr><td rowspan="4">F_A</td><td>A_1</td><td>0.027</td><td>0.054</td></tr>
<tr><td>B_1</td><td>0.046</td><td>0.092</td></tr>
<tr><td>C_1</td><td>0.048</td><td>0.096</td></tr>
<tr><td>D_1</td><td>0.022</td><td>0.044</td></tr>
<tr><td rowspan="4">F_B</td><td>A_2</td><td>0.054</td><td>0.081</td></tr>
<tr><td>B_2</td><td>0.092</td><td>0.138</td></tr>
<tr><td>C_2</td><td>0.096</td><td>0.144</td></tr>
<tr><td>D_2</td><td>0.044</td><td>0.066</td></tr>
</table>

Table 6.6. *Parameters of the three models*

We use the following notation:

– Volume(m_A, m_B, m_C, m_D) is the volume of the wing;

– $f_a(A_1, B_1, C_1, D_1)$ is the lower bound of the critical frequency band;

– $f_b(A_2, B_2, C_2, D_2)$ is the upper bound of the critical frequency band;

– $d_{\beta_i}(.)$ represents the image of the reliability index in the hybrid space.

SP method

In this section, the formulation of the problem is simplified, so we have two optimization problems:

– the first minimizes the objective function of the first model subject to the constraint associated with the frequency f_a:

$$\begin{cases} \min\limits_{A_1,\ldots} \text{Vol}_a\,(A_1, B_1, C_1, D_1) \\ \text{s.t.} f_{\max}(A_1, B_1, C_1, D_1) - f_a = 0 \end{cases} \qquad [6.41]$$

– the second minimizes the objective function of the second model subject to the constraint associated with the frequency f_b:

$$\begin{cases} \min\limits_{A_2,\ldots} \mathrm{Vol}_b\,(A_2, B_2, C_2, D_2) \\ \text{s.t.} f_{\max}(A_2, B_2, C_2, D_2) - f_b = 0 \end{cases} \quad [6.42]$$

– after the optimization problems, we calculate the coordinates of the optimal point corresponding to the third model subject to the frequency f_n. The average solution of the safest point is given by a point in the interval $[f_a, f_b]$.

For a normal distribution, we have that:

$$m_i = x_i = \frac{y_i^a + y_i^b}{2}, \qquad i = 1, \ldots, n \quad [6.43]$$

For a log-normal distribution, we have that:

$$m_i = x_i = \sqrt{1 + \gamma_i^2}\exp\left(\frac{\ln(y_i^a \cdot y_i^b)}{2}\right), \qquad i = 1, \ldots, n \quad [6.44]$$

Numerical results

The two tables given below show the results of RBDO with both methods: hybrid and SP. We present the results obtained using two different distributions to demonstrate the effectiveness of the proposed new method.

For a normal distribution

Table 6.7 presents the results of RBDO with two methods, hybrid and SP, in the case of a normal distribution.

With a normal distribution, the objective is to provide the best compromise between cost and design safety. This objective can easily be achieved with the new SP method, which produces a better solution than the hybrid method. The optimal design point obtained by the SP method satisfies the target reliability level of $\beta_t = 3$ exactly. The hybrid method on the other hand results in a reliability level of $\beta_t = 3.0153$ (i.e. within the small tolerance margin of 0.51%).

Variables	Parameters	Initial	Optimal design with SP	Optimal design with HM
F_N	A_N	0.054	0.040826	0.057710
	B_N	0.092	0.069145	0.084087
	C_N	0.096	0.095406	0.093940
	D_N	0.044	0.037493	0.037192
F_A	A_1	0.027	0.027285	0.045907
	B_1	0.046	0.046169	0.071424
	C_1	0.048	0.082863	0.080665
	D_1	0.022	0.028582	0.032862
F_B	A_2	0.081	0.054367	0.067262
	B_2	0.138	0.092121	0.094616
	C_2	0.144	0.10795	0.10657
	D_2	0.066	0.046404	0.045188
State variables	F_A	23.746	24.998	25.810
	F_B	46.788	34.003	34.069
	F_N	31.414	29.102	30.057
	MASS	411.310	359.855	387.794
	Time (s)	–	1100	15270

Table 6.7. *Results for the normal distribution*

The volume of the aircraft wing is slightly better with the hybrid method than with the SP method, but the numerical cost of the hybrid method is much higher than that of the SP method. Figures 6.23 and 6.24 show the change in the objective function and the design variables as a function of the iterations of the hybrid method.

Log-normal distribution

Table 6.8 presents the results of RBDO with two methods, hybrid and SP, in the case of a log-normal distribution. Figures 6.25 and 6.26 show the change in the objective function and the design variables as a function of the iterations of the hybrid method.

For log-normal distributions, the hybrid and the SP methods satisfy the target reliability level $\beta_t = 3$ exactly, achieving essentially equivalent optimal volumes, but the hybrid requires higher computation times.

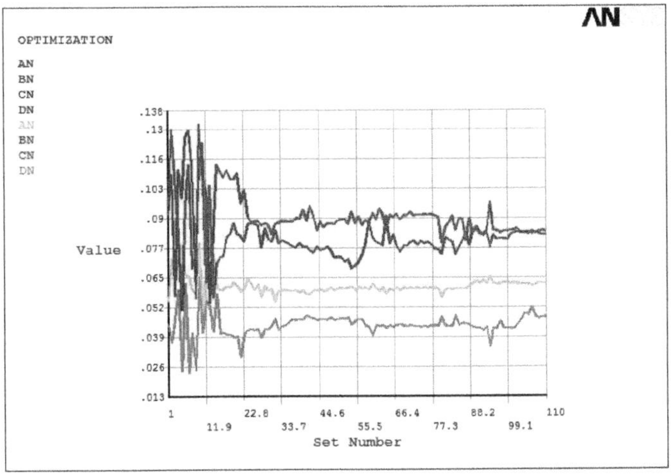

Figure 6.23. *Change in the design variables as a function of the iterations. For a color version of this figure, see www.iste.co.uk/elhami/interactions.zip*

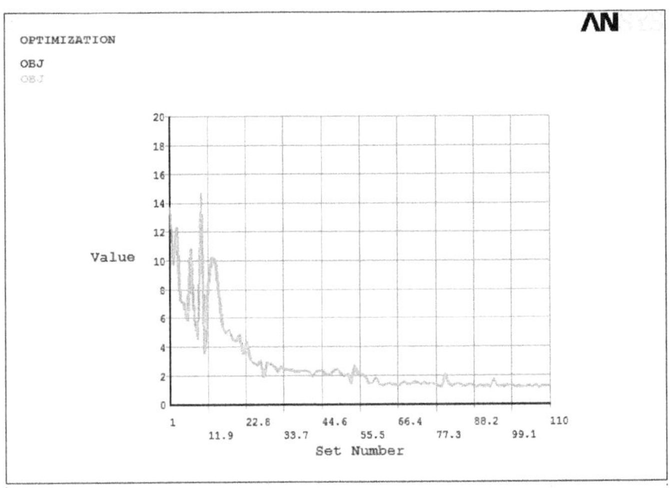

Figure 6.24. *Change in the objective function as a function of the iterations. For a color version of this figure, see www.iste.co.uk/elhami/interactions.zip*

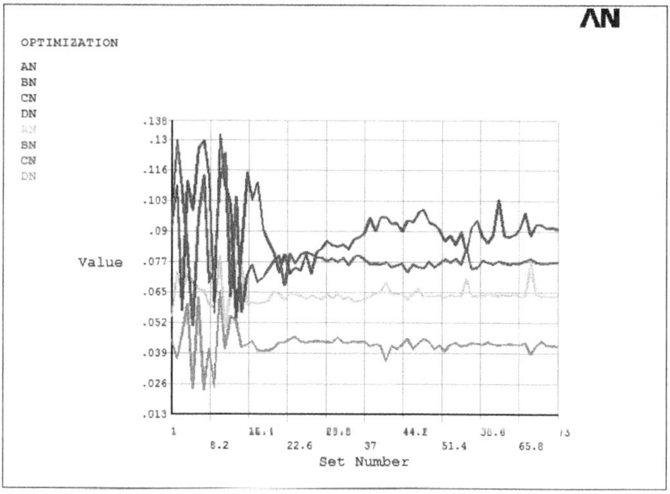

Figure 6.25. *Change in the design variables as a function of the iterations. For a color version of this figure, see www.iste.co.uk/elhami/interactions.zip*

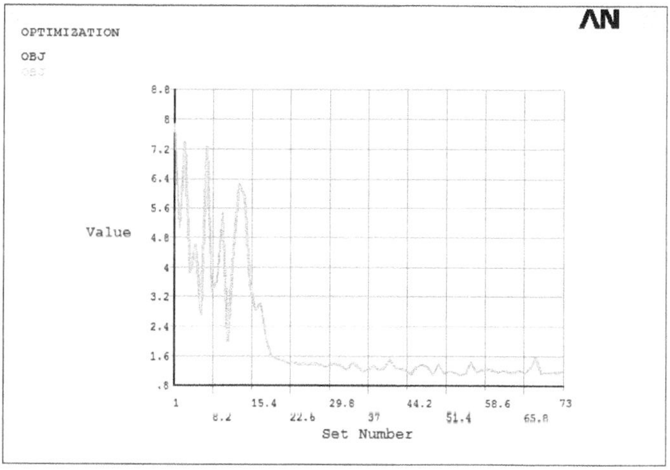

Figure 6.26. *Change in the objective function as a function of the iterations. For a color version of this figure, see www.iste.co.uk/elhami/interactions.zip*

Variables	Parameters	Initial	Optimal design with SP	Optimal design with HM
F_N	A_N	0.054	0.038582	0.054411
	B_N	0.092	0.070540	0.087850
	C_N	0.096	0.095833	0.079661
	D_N	0.044	0.035276	0.044905
F_A	A_1	0.027	0.027285	0.037127
	B_1	0.046	0.046169	0.087578
	C_1	0.048	0.082863	0.071878
	D_1	0.022	0.028582	0.037519
F_B	A_2	0.081	0.054367	0.071352
	B_2	0.138	0.092121	0.092247
	C_2	0.144	0.10795	0.10706
	D_2	0.066	0.046404	0.046619
State variables	F_A	23.746	24.998	25.95
	F_B	46.788	34.003	34.11
	F_N	31.414	29.102	28
	MASS (kg)	411.310	356.410	387.78
	Time (s)	-	1100	16510

Table 6.8. *Results for the log-normal distribution*

Bibliography

[AIT 97] AIT YOUNES T., HAMDI M., "Modal shapes reconstruction method for large domains (structure and acoustic)", *Computational Acoustics and its Environmental Applications II*, Computational Mechanics Publications, ASME, 1997.

[ALL 02] ALLERY C., Contribution à l'identification des bifurcations et à l'étude des écoulements fluides par des systèmes d'ordre faible (POD), PhD Thesis, University of Poitiers, France, 2002.

[ANG 99] ANGOT P., BRUNEAU C.H., FABRIE P., "A penalization method to take into account obstacles in incompressible viscous flows", *Numerische Mathematik*, vol. 81, no. 4, pp. 497–520, 1999.

[ANS 12a] ANSYS INC., Ansys Workbench 14.0- User Guide, 2012.

[ANS 12b] ANSYS INC., Ansys Mechanical APDL Coupled-Field Analysis Guide, 2012.

[ANS 13] ANSYS INC., Ansys Workbench 14.0- User Guide, ANSYS, Inc. October outhpointe 275 Technology Drive Canonsburg, PA 15317, 2013.

[ANS 15] ANSYS INC., Ansys Mechanical APDL, 2015.

[API 03] APIWATTANNALUNGGARN P., Model reduction of nonlinear structural systems using nonlinear normal modes and component mode synthesis, PhD Thesis, Michigan State University, 2003.

[AQU 04] AQUELET N., Modélisation de l'impact hydrodynamique par un couplage fluide-structure, PhD Thesis, Lille University of Science and Technologies, 2004.

[AUB 88] AUBRY N., HOLMES P., LUMLEY J. *et al.*, "The dynamics of coherent structures in the wall region of a turbulent boundary layer", *Journal of Fluid Mechanics*, vol. 192, pp. 115–173, 1988.

[AXI 01] AXISA F., *Modélisation des systèmes mécaniques. Interactions fluide structure*, Hermès Science, Paris, 2001.

[AZR 99] AZRAR L., BENAMAR R., WHITE R., "A semi-analytical approach to the non-linear dynamic response problem of s-s and c-c beams at large vibration amplitudes, Part I: general theory and application to the single mode approach to free and forced vibration analysis", *Journal of Sound and Vibration*, vol. 224, no. 2, pp. 183–207, 1999.

[BAT 82] BATHE K., *Finite Element Procedures in Engineering Analysis*, Prentice-Hall, New Jersey, 1982.

[BAT 90] BATOZ J.-L., DHATT G., *Modélisation des structures par éléments finis*, Hermès Science, Paris, 1990.

[BEL 00] BELYTSCHKO T., LIU W., MORAND B., *Non-linear Finite Element for Continua and Structure*, John Wiley & Sons, New York, 2000.

[BEN 71] BENFIELD W.A., HRUDA R.F., "Vibration analysis of structures by component mode synthesis", *AIAA Journal*, vol. 9, no. 7, pp. 1255–1261, 1971.

[BEN 99] BENNIGHOF J., "Vibroacoustic frequency sweep analysis using automated multi-level substructuring", *AIAA Journal*, vol. 1, pp. 422–427, 1999.

[BEN 07] BENAOUICHA M., Contribution au développement d'algorithmes de couplage en interaction fluide-structure, PhD Thesis, University of La Rochelle, 2007.

[BEN 09] BENDAOU O., ROJAS J., EL HAMI A. *et al.*, "Stochastic and reliability analysis of a propeller with model reduction", *European Journal of Computational Mechanics*, vol. 18, no. 2, pp. 153–173, 2009.

[BER 99] BERMUDEZ A., HERVELLA-NIETO L., RODRIGUEZ R., "Finite element computation of three-dimensional elastoacoustic vibrations", *Journal of Sound and Vibration*, vol. 219, pp. 279–306, 1999.

[BER 04] BERTOLUZZA S., ISMAIL M., MAURY B., *The FBM method: semi-discrete scheme and some numerical experiments*, Springer-Verlag, 2004.

[BIR 07] BIRD R., STEWART W., LIGHTFOOT N., *Transport phenomena*, 2nd ed., John Wiley & Sons, New York, 2007.

[BOF 03] BOFFI D., GASTALDI L., "A finite element approach for the immersed boundary method", *Computers Structures*, vol. 81, pp. 491–501, 2003.

[BOR 06] BORZA D., MAKHLOUFI A., EL HAMI A., "Holographic vibration measurement and numerical modelling of immersed structures", *Proceedings of Speckle'06*, Nimes, vol. 6341, 2006.

[BOU 87] BOUJOT J., "Mathematical formulation of fluid-structure interaction problems", *Mathematical Modeling and Numerical Analysis*, vol. 21, pp. 239–260, 1987.

[BRE 89] BREITUNG K., HOHENBICHLER M., "Asymptotic approximations for multivariate integrals with an application to multinormal probabilities", *Journal of Multivariate Analysis*, vol. 30, no. 1, pp. 80–97, 1989.

[CAU 04] CAUSIN P., GERBEAU J.F., NOBILE F., Added-mass effect in the design of partitioned algorithms for fluid-structure problems, Research Report, INRIA, 2004.

[CAZ 98] CAZEMIER W., VERSTAPPEN R., VELDMAN A., "Proper orthogonal decomposition and low-dimensional models for driven cavity", *Physics of Fluids*, vol. 10, no. 7, pp. 1685–1699, 1998.

[CRA 68] CRAIG R.R., BAMPTON M.C., "Coupling of substructures for dynamic analyses", *AIAA Journal*, vol. 6, pp. 1313–1319, 1968.

[CRA 76] CRAIG R.R., CHANG C.J., "Free-interface methods of substructures coupling for dynamic analysis", *AIAA Journal*, vol. 14, no. 11, 1976.

[CRA 95] CRAIG R.R., "Substructure methods in vibration", *Journal of Vibration and Acoustics*, vol. 117, pp. 207–213, 1995.

[DEL 01] DE-LANGRE E., *Fluides et Solides*, Les éditions de l'Ecole polytechnique, Paris, 2001

[DER 95] DERMIDZIC I., MUZAFERIJA S., "Numerical method for coupled fluid flow, heat transfer and stress analysis using unstructured moving meshes with cells of arbitrary topology", *Computer Methods in Applied Mechanics and Engineering*, vol. 125, pp. 235–255, 1995.

[DES 08] DESTUYNDER P., *Modélisation en mécanique des fluides et des structures : aéroélasticité et aéroacoustique*, Hermès Science-Lavoisier, Paris, 2008.

[DEV 02] DEVILLE M.O., F.F.P., MUND E.H., *High-Order Methods for Incompressible Fluid Flow*, Cambridge University Press, 2002.

[DHA 04] DHATT G., TOUZOT G., LEFRANCOIS E., *Méthode des éléments finis*, Hermès Science-Lavoisier, Paris, France, 2004.

[DIN 07] DINIZ DOS SANTOS N.M., Numerical methods for fluid-structure interaction problems with valves, PhD Thesis, University Pierre and Marie Curie, Paris VI, 2007.

[DON 82] DONEA J., GUILIANI S., HALLEUX J.P., "An arbitrary Lagrangian-Eulerian finite element method for transient dynamic fluid structure interactions", *Computer Methods in Applied Mechanics and Engineering*, vol. 33, pp. 689–723, 1982.

[DON 03] DONEA J., HUERTA H., *Finite element methods for flow problems*, John Wiley & Sons, 2003.

[ELH 13a] EL HAMI A., RADI B., *Incertitudes, optimisation et fiabilité des structures*, Hermès Science-Lavoisier, Paris, 2013.

[ELH 13b] EL HAMI A., RADI B., *Uncertainty and Optimization in Structural*, John Wiley and Sons, 2013.

[ELH 15] EL HAMI A., RADI B., "Optimisation et fiabilité des systèmes complexes", *Techniques de l'ingénieur*, no. MT9313 v1, pp. 1–17, 2015.

[ELM 15] EL MAANI R., RADI B., EL HAMI A., "Reliability study of a coupled three-dimensional system with uncertain parameters", *Advances in Theoretical and Applied Mechanics*, vol. 8, pp. 87–93, 2015.

[EVE 81] EVERSTINE G.C., "A symmetric potential formulation for fluid-structure interaction", *Journal of Sound and Vibration*, vol. 79, pp. 157–160, 1981.

[FAR 98] FARHAT C., DEGAND C., KOOBUS B., LESOINNE M., "Torsional springs for two-dimensional dynamic unstructured fluid meshes", *Computer Methods in Applied Mechanics and Engineering*, vol. 163, pp. 231–245, 1998.

[FAR 00] FARHAT C., LESOINNE M., "Two efficient staggered algorithms for the serial and parallel solution of three-dimensional nonlinear transient aeroelastic problems", *Computer Methods in Applied Mechanics and Engineering*, vol. 182, pp. 499–515, 2000.

[FEN 86] FENG Y., MOSES F., "A method of structural optimization based on structural system reliability", *Journal of Structural Mechanics*, vol. 14, pp. 437–453, 1986.

[FER 99] FERREIRA J.V., Dynamic response analysis of structures with nonlinear components, PhD Thesis, Imperial College of Science, Technology and Medecine, University of London, May 1999.

[FER 05] FERNÁNDEZ M.A., GERBEAU J.F., GRANDMONT C., A projection semi-implicite scheme for the coupling of an elastic structre with an incompressible fluid, Research Report, INRIA, 2005.

[FER 06] FERNÁNDEZ M.A., GERBEAU J.F., GRANDMONT C., "A projection algorithm for fluid/structure interaction problems with strong added-mass effect", *Comptes Rendus Mathématique*, vol. 342, pp. 279–284, 2006.

[FOR 04] FORMAGGIA L., NOBILE F., "Stability analysis of second-order time accurate schemes for ALE-FEM", *Computer Methods in Applied Mechanics and Engineering*, vol. 193, pp. 4097–4116, 2004.

[FRA 94] FRANGOPOL D., HENDAWI S., "Incorporation of corrosion effects in reliability-based optimization of composite hybrid plate girders", *Structural Safety*, vol. 16, pp. 145–169, 1994.

[GAU 11] GAUGAIN F., Analyse expérimentale et simulation numérique de l'interaction fluide-structure d'un hydrofoil élastique en écoulement subcavitant et cavitant, PhD Thesis, École Nationale Supérieure d'Arts et Métiers, Paris, 2011.

[GER 03] GERBEAU J.F., VIDRASCU M., A quasi-Newton algorithm based on a reduced model for fluid/structure interaction problems in blood flows, Research Report, INRIA, 2003.

[GIB 86] GIBERT R., *Vibration des structures. Intéraction avec les fluides. Sources d'exitation aléatoires*, Eyrolles, Paris, 1986.

[GIR 03] GIRARD A., ROY N., *Dynamique des structures industrielles*, Hermès Science-Lavoisier, Paris, 2003.

[GLO 94] GLOWINSKI R., PAN T.W., PERIAUX J., "A fictitious domain method for Dirichlet problem and applications", *Computer Methods in Applied Mechanics and Engineering*, vol. 111, pp. 283–303, 1994.

[GRA 12] GRANDMONT C., MADAY Y., "Fluid-Structure Interaction: a theoretical point of view", *Revue Européenne des Éléments Finis*, vol. 9, nos. 6–7, pp. 633–653, 2012.

[GUE 99] GUERICH M., HAMDI M., "A numerical method for vibro-acoustic problems with incompatible finite element meshes using b-Spline functions", *Journal of the Acoustical Society of America*, vol. 105, no. 1, pp. 1682–1694, 1999.

[GUE 06] GUERMOND J., MINEV P., SHEN J., "An overview of projection methods for incompressible flows", *Computer Methods in Applied Mechanics and Engineering*, vol. 195, nos. 44–47, pp. 6011–6045, 2006.

[HÜB 04] HÜBNER B., WALHORN E., DINKLER D., "A monolithic approach to fluid-structure interaction using space time finite elements", *Computer Methods in Applied Mechanics and Engineering*, vol. 193, pp. 2087–2104, 2004.

[HAM 78] HAMDI M., OUSSET Y., VERCHERY G., "A displacement method for the analysis of vibrations of coupled fluid-structure systems", *International journal of numerical methods in engineering*, vol. 13, pp. 139–150, 1978.

[HIR 74] HIRT C., AMSDEN A., COOK J., "An arbitrary Lagrangian-Eulerian computing method for all flow speeds", *Journal of Computational Physics*, vol. 14, no. 3, pp. 227–253, 1974.

[HOU 12] HOU G., WANG J., LAYTON A., "Numerical methods for fluid-structure interaction", *Communications in Computational Physics*, vol. 12, no. 2, pp. 337–377, 2012.

[HU 01] HU H.H., PATANKAR N., ZHU M., "Direct numerical simulations of Fluid-Solid systems using the arbitrary Lagrangian-Eulerian technique", *Journal of Computational Physics*, vol. 169, no. 2, pp. 427–462, 2001.

[HUR 65] HURTY W., "Vibration of structural systems by component mode synthesis", *AIAA Journal*, vol. 6, no. 7, pp. 678–685, 1965.

[IWA 03] IWATSUBO T., SHIMBO K., KAWAMURA S., "Nonlinear vibration analysis of a rotor system using component mode synthesis method", *Archive of Applied Mechanics*, vol. 72, no. 11–12, 2003.

[KHA 02] KHARMANDA G., MOHAMED A., LEMAIRE M., "Efficient reliability-based design optimization using a hybrid space with application to finite element analysis", *Structural and Multidisciplinary Optimization*, vol. 24, no. 3, pp. 233–245, 2002.

[LEG 01] LEGRESLEY P., ALONSO J., "Investigation of non-linear projection for pod based reduced order models for aerodynamics", *AIAA Journal*, vol. 42, no. 8, pp. 9–26, 2001.

[LEV 94] LEVEQUE R.J., LI Z., "The immersed interface method for elliptic equations with discontinuous coefficients and singular sources", *Journal on Numerical analysis*, vol. 31, no. 4, pp. 1019–1044, 1994.

[LIA 02] LIANG Y., LEE H., LIM S. *et al.*, "Proper orthogonal decomposition and its applications-part I: theory", *Journal of Sound and Vibration*, vol. 252, no. 3, pp. 527–544, 2002.

[LIA 03] LIAN Y., SHYY W., VIIERU D. *et al.*, "Membrane Wing aerodynamics for micro air Vehicles", *Progress in Aerospace Sciences*, vol. 39, no. 6–7, pp. 467–510, 2003.

[LUM 67] LUMLEY J., "The structure of inhomogeneous turbulent flows", *Atmospheric Turbulence and Radio Wave Propagation, Intern. Collq.*, Moscow, Russia, pp. 166–178, 1967.

[MAC 71] MACNEAL R., "A hybrid method of component mode synthesis", *Computers and Structures*, vol. 1, pp. 581–601, 1971.

[MAK 08] MAKHLOUFI A., Contribution aux développements et à la validation des méthodes de l'optimisation fiabiliste de conception en dynamique des structures mécaniques, Thesis, INSA de Rouen, 2008.

[MAM 95] MAMAN N., FARHAT C., "Matching fluid and structure meshes for aeroelastic computations: a parallel approach", *Computers and Structures*, vol. 54, no. 4, pp. 779–785, 1995.

[MAN 13a] MANSOURI M., RADI B., EL HAMI A., "Reliability analysis of vibro-acoustic problem", *Advanced Materials Research*, vol. 682, pp. 73–83, 2013.

[MAN 13b] MANSOURI M., RADI B., EL HAMI A., "Contribution to the analysis and numerical simulation of vibro-acoustic phenomena by the modal synthesis methods", *Advances in Theoretical and Applied Mechanics*, vol. 6, no. 1, pp. 1–12, 2013.

[MAR 78] MARCUS M., "A finite element method applied to the vibration of submersed plates", *Journal of Ship Research*, vol. 22, pp. 94–99, 1978.

[MAU 99] MAURY B., "Direct simulations of 2D fluid particle flows in biperiodic domains", *Journal of Computational Physics*, vol. 156, no. 2, pp. 325–351, 1999.

[MIT 05] MITTAL R., IACCARINO G., "Immersed boundary methods", *Annual Review of Fluid Mechanics*, vol. 37, no. 1, pp. 239–261, 2005.

[MOH 06] MOHSINE A., Contribution à l'optimisation fiabiliste en dynamique des structures mécaniques, Thesis, INSA de Rouen, 2006.

[MOR 79] MORAND H., OHAYON R., "Substructure variational analysis of the vibration of coupled fluid-structure systems. Finite element results", *International Journal for Numerical Methods in Engineering*, vol. 14, pp. 741–755, 1979.

[MOR 95] MORAND J., OHAYON R., *Fluid-structure Interaction: Applied Numerical Methods*, John Wiley & Sons, 1995.

[MOR 12] MORINISHI K., FUKUI T., "An Eulerian approach for fluid-structure interaction problems", *Computers and Fluids*, vol. 35, pp. 92–98, 2012.

[NEW 96] NEWMAN A., Model reduction via the Karhunen-Loeve expansion part i: an exposition, Technical report, Inst. Systems Research, 1996.

[NGU 05] NGUYEN T., ARGOUL P., BONNET G. *et al.*, "Analyse dynamique d'une poutre linéaire avec une interface non-linéaire", *7e Colloque National en Calcul des Structures*, pp. 345–350, 2005.

[OLS 83] OLSON L., BATHE K., "A study of displacement-based fluid finite elements for calculating frequencies of fluid and fluid-structure systems", *Nuclear Engineering and Design*, vol. 76, pp. 137–151, 1983.

[OLS 85] OLSON L., BATHE K., "Analysis of fluid-structure interactions. A direct symmetric coupled formulation based on the fluid velocity potential", *Computers and Structures*, vol. 21, pp. 21–32, 1985.

[OMU 99] OMURTAG A., SIROVICH L., "On low-dimensional modeling of channel turbulence", *Theoretical and Computational Fluid Dynamics*, vol. 13, pp. 115–127, 1999.

[PAT 80] PATANKAR S., *Numerical Heat Transfer and Fluid Flow*, McGraw Hill–CRC Press, 1980.

[PAT 00] PATANKAR N., SINGH P., JOSEPH D. *et al.*, "A new formulation of the distributed lagrange multiplier/fictitious domain method for particulate flows", *International Journal of Multiphase Flow*, vol. 26, no. 9, pp. 1509–1524, 2000.

[PAT 01] PATANKAR N., A formulation for fast computations of rigid particulate flows, Report, Annual Research Briefs (Center for turbulence research), 2001.

[PES 77] PESKIN C., "Numerical analysis of blood flow in the heart", *Journal of Computational Physics*, vol. 25, no. 3, pp. 220–252, 1977.

[PES 03] PESKIN C., "The immersed boundary method", *Acta Numerica*, vol. 11, pp. 479–517, 2003.

[PIP 95] PIPERNO S., FARHAT C., LARROUTUROU B., "Partitioned procedures for the transient solution of coupled aroelastic problems Part I : model problem, theory and two dimensional application", *Computer Methods in Applied Mechanics and Engineering*, vol. 124, nos.1–2, pp. 79–112, 1995.

[PIP 97] PIPERNO S., "Explicit-implicit fluid-structure staggered procedures with a structural predictor and fluid subcycling for 2D inviscid aeroelastic simulations", *International Journal for Numerical Methods in Fluids*, vol. 25, pp. 1207 1226, 1997.

[PIP 00] PIPERNO S., FARHAT C., "Design of efficient partitioned procedures for the transient solution of aeroelastic problem", *Revue Européenne des Eléments Finis*, vol. 9, nos. 6–7, pp. 655–680, 2000.

[PIP 01] PIPERNO S., FARHAT C., "Partitioned procedures for the transient solution of coupled aroelastic problems – Part II : energy transfer analysis and three-dimensional applications", *Computer Methods in Applied Mechanics and Engineering*, vol. 190, nos. 24–25, pp. 3147–3170, 2001.

[PIR 89] PIRONNEAU O., *Finite element method for fluids*, John Wiley & Sons, New York, 1989.

[PON 09] PONCET P., "Analysis of an immersed boundary method for three-dimensional flows in vorticity formulation", *Journal of Computational Physics*, vol. 228, no. 19, pp. 7268–7288, 2009.

[QUA 94] QUARTERONI A., VALLI A., *Numerical Approximation of Partial Differential Equations*, Springer-Verlag, Berlin, Heidelberg, 1994.

[RAD 15] RADI B., EL HAMI A., "RBDO analysis of a vibro-acoustique system: application to a marine propeller", *Structural Nonlinear Dynamics and Diagnosis*, Springer International Publishing, 2015.

[REM 94] REMPFER D., FASEL H., "Evolution of three-dimensional coherent structures in a transitional flat-plate boundary layer", *Journal of Fluid Mechanics*, vol. 275, pp. 257–283, 1994.

[REM 96] REMPFER D., "Investigations of boundary layer transition via galerkin projections on empirical eigenfunctions", *Physics of Fluids*, vol. 8, pp. 175–188, 1996.

[RUB 75] RUBIN S., "Improved component-mode representation for structural dynamic analysis", *AIAA Journal*, vol. 13, pp. 995–1006, 1975.

[SAN 88] SANDBERG G., GORANSSON P., "A symmetric finite element formulation for acoustic fluid-structure interaction analysis", *Journal of Sound and Vibration*, vol. 123, no. 3, pp. 507–515, 1988.

[SAN 01] SANDBERG G., HANSSON P., GUSTAVSSON M., "Domain decomposition in acoustic and structure-acoustic analysis", *Computer Methods in Applied Mechanics and Engineering*, vol. 190, pp. 2979–2988, 2001.

[SCH 74] SCHMIT L., FARSHI B., "Some approximation concepts for structural synthesis", *AIAA Journal*, vol. 12, no. 3, pp. 692–699, 1974.

[SCH 06] SCHAEFER M., *Computational Engineering–Introduction to Numerical Methods*, Springer-Verlag, Berlin, Heidelberg, 2006.

[SHA 93] SHAW S., PIERRE C., "Normal modes of vibration for non-linear vibratory systems", *Journal of Sound and Vibration*, vol. 164, no. 1, pp. 85–124, 1993.

[SHA 05] SHARMA N., PATANKAR N., "A fast computation technique for the direct numerical simulation of rigid particulate flows", *Journal of Computational Physics*, vol. 205, no. 2, pp. 439–457, 2005.

[SIG 07] SIGRIST J., LAINE C., PESEUX B., "Dynamic analysis of fluid-structure interaction problems with spectral method using pressure-based finite elements", *Finite Element Analysis in Design*, Elsevier Science Publishers, Amsterdam, 2007.

[SIG 10] SIGRIST J., Méthodes numériques de calculs couplés fluide/structure : Cas du fluide s'écoulant, Report, Techniques de l'Ingénieur, 2010.

[SIG 11] SIGRIST J., *Intéraction fluide-structure, Analyse vibratoire par éléments finis*, Ellipses, 2011.

[SIR 87] SIROVITCH L., "Turbulence and the dynamics of coherent structures, parti: Coherent strucures, partii: symmetries and transformations, part iii: dynamics and scaling", *Quarterly of Applied Mathematics*, vol. 45, no. 3, pp. 561–590, 1987.

[SLA 02] SLATER J., ABBOTT J., CAVICCHI R., "Validation of Wind for a series of Inlet Flows", *AIAA Journal*, vol. 42, pp. 1725–1732, 2002.

[SOU 09] SOULI M., SIGRIST J. (eds.), *Intéraction fluide-structure: modélisation et simulation numérique*, Hermès Science-Lavoisier, Paris, 2009.

[SOU 10] SOULI M., BENSON D. (eds.), *Arbitrary Lagrangian-Eulerian and Fluid-Structure Interaction*, ISTE, London and John Wiley & Sons, New York, 2010.

[SUN 98] SUNDARARAJAN P., NOAH S., "An algorithm for response and stability of large order nonlinear systems – application to rotor systems", *Journal of Sound and Vibration*, vol. 214, no. 4, pp. 695–723, 1998.

[TRA 01] TRAN D., "Component mode synthesis methods using interface modes. Application to structures with cyclic symmetry", *Computers and Structures*, vol. 79, pp. 209–222, 2001.

[VUI 11] VUIK C., SEGAL A., Iterative solution methods and finite element methods for the incompressible Navier-Stokes equations, Technical Report, Delft University of Technology, 2011.

[WAN 97] WANG X., BATHE K.J., "Displacement/pressure based mixed finite element formulations for acoustic fluid-structure interaction problems", *International Journal for Numerical Methods in Engineering*, vol. 40, pp. 2001–2017, 1997.

[WU 03] WU G., LIANG Y., LIN W. *et al.*, "A note on equivalence of proper orthogonal decomposition methods", *Journal of Sound and Vibration*, vol. 265, no. 5, 2003.

[XIN 96] XING J., PRICE W., DU Q., "Mixed finite element substructure-subdomain methods for the dynamical analysis of coupled fluid-solid interaction problems", *Philosophical Transactions of the Royal Society London, Series A Mathematical, Physical and Engineering Sciences*, vol. 354, pp. 259–295, 1996.

[ZHA 07] ZHANG L., GAY M., "Immersed finite element method for fluid-structure interactions", *Journal of Fluids and Structures*, vol. 23, pp. 839–857, 2007.

[ZIE 78] ZIENKIEWICZ O., BETTESS P., "Fluid-structure dynamic interaction and wave forces. An introduction to numerical treatment", *International Journal for Numerical Methods in Engineering*, vol. 13, pp. 1–16, 1978.

[ZIE 89] ZIENKIEWICZ O., TAYLOR R., *The Finite Element Method. Basic Formulation and Linear Problems*, McGraw-Hill, Butterworth, Heinemann, 1989.

Index

R, S

T, V, W

Other titles from

iSTE

in

Mechanical Engineering and Solid Mechanics

2016

BOREL Michel, VÉNIZÉLOS Georges
Movement Equations 1: Location, Kinematics and Kinetics
(Non-deformable Solid Mechanics Set – Volume 1)

Movement Equations 2: Mathematical and Methodological Supplements
(Non-deformable Solid Mechanics Set – Volume 2)

BOYARD Nicolas
Heat Transfer in Polymer Composite Materials

CARDON Alain, ITMI Mhamed
New Autonomous Systems
(Reliability of Multiphysical Systems Set – Volume 1)

DAHOO Pierre Richard, POUGNET Philippe, EL HAMI Abdelkhalak
Nanometer-scale Defect Detection Using Polarized Light
(Reliability of Multiphysical Systems Set – Volume 2)

DE SAXCÉ Géry, VALLÉE Claude
Galilean Mechanics and Thermodynamics of Continua

DORMIEUX Luc, KONDO Djimédo
Micromechanics of Fracture and Damage (Micromechanics Set – Volume 1)

EL HAMI Abdelkhalak, RADI Bouchaib
Stochastic Dynamics of Structures (Mathematical and Mechanical Engineering Set – Volume 2)

GOURIVEAU Rafael, MEDJAHER Kamal, ZERHOUNI Noureddine
From Prognostics and Health Systems Management to Predictive Maintenance 1: Monitoring and Prognostics (Reliability of Multiphysical Systems Set – Volume 4)

KHARMANDA Ghias, EL HAMI Abdelkhalak
Reliability in Biomechanics (Reliability of Multiphysical Systems Set – Volume 3)

Biomechanics: Optimization, Uncertainties and Reliability (Reliability of Multiphysical Systems Set – Volume 5)

LEDOUX Michel, EL HAMI Abdelkhalak
Fluid Mechanics: Analytical Methods(Mathematical and Mechanical Engineering Set – Volume 3)

Compressible Flow Propulsion and Digital Approaches in Fluid Mechanics (Mathematical and Mechanical Engineering Set – Volume 4)

MOLIMARD Jérôme
Experimental Mechanics of Solids and Structures

MORI Yvon
Mechanical Vibrations: Applications to Equipment

RADI Bouchaib, EL HAMI Abdelkhalak
Material Forming Processes: Simulation, Drawing, Hydroforming and Additive Manufacturing (Mathematical and Mechanical Engineering Set – Volume 1)

2015

KARLIČIĆ Danilo, MURMU Tony, ADHIKARI Sondipon, MCCARTHY Michael
Non-local Structural Mechanics

SAB Karam, LEBÉE Arthur
Homogenization of Heterogeneous Thin and Thick Plates

2014

ATANACKOVIC M. Teodor, PILIPOVIC Stevan, STANKOVIC Bogoljub, ZORICA Dusan
Fractional Calculus with Applications in Mechanics: Vibrations and Diffusion Processes

ATANACKOVIC M. Teodor, PILIPOVIC Stevan, STANKOVIC Bogoljub, ZORICA Dusan
Fractional Calculus with Applications in Mechanics: Wave Propagation, Impact and Variational Principles

CIBLAC Thierry, MOREL Jean-Claude
Sustainable Masonry: Stability and Behavior of Structures

ILANKO Sinniah, MONTERRUBIO Luis E., MOCHIDA Yusuke
The Rayleigh–Ritz Method for Structural Analysis

LALANNE Christian
Mechanical Vibration and Shock Analysis – 5-volume series – 3rd edition
Sinusoidal Vibration – volume 1
Mechanical Shock – volume 2
Random Vibration – volume 3
Fatigue Damage – volume 4
Specification Development – volume 5

LEMAIRE Maurice
Uncertainty and Mechanics

2013

ADHIKARI Sondipon
Structural Dynamic Analysis with Generalized Damping Models: Analysis

ADHIKARI Sondipon
Structural Dynamic Analysis with Generalized Damping Models: Identification

BAILLY Patrice
Materials and Structures under Shock and Impact

BASTIEN Jérôme, BERNARDIN Frédéric, LAMARQUE Claude-Henri
Non-smooth Deterministic or Stochastic Discrete Dynamical Systems: Applications to Models with Friction or Impact

EL HAMI Abdelkhalak, BOUCHAIB Radi
Uncertainty and Optimization in Structural Mechanics

KIRILLOV Oleg N., PELINOVSKY Dmitry E.
Nonlinear Physical Systems: Spectral Analysis, Stability and Bifurcations

LUONGO Angelo, ZULLI Daniele
Mathematical Models of Beams and Cables

SALENÇON Jean
Yield Design

2012

DAVIM J. Paulo
Mechanical Engineering Education

DUPEUX Michel, BRACCINI Muriel
Mechanics of Solid Interfaces

ELISHAKOFF Isaac *et al.*
Carbon Nanotubes and Nanosensors: Vibration, Buckling and Ballistic Impact

GRÉDIAC Michel, HILD François
Full-Field Measurements and Identification in Solid Mechanics

GROUS Ammar
Fracture Mechanics – 3-volume series
Analysis of Reliability and Quality Control – volume 1
Applied Reliability – volume 2
Applied Quality Control – volume 3

RECHO Naman
Fracture Mechanics and Crack Growth

2011

KRYSINSKI Tomasz, MALBURET François
Mechanical Instability

SOUSTELLE Michel
An Introduction to Chemical Kinetics

2010

BREITKOPF Piotr, FILOMENO COELHO Rajan
Multidisciplinary Design Optimization in Computational Mechanics

DAVIM J. Paulo
Biotribolgy

PAULTRE Patrick
Dynamics of Structures

SOUSTELLE Michel
Handbook of Heterogenous Kinetics

2009

BERLIOZ Alain, TROMPETTE Philippe
Solid Mechanics using the Finite Element Method

LEMAIRE Maurice
Structural Reliability

2007

GIRARD Alain, ROY Nicolas
Structural Dynamics in Industry

GUINEBRETIÈRE René
X-ray Diffraction by Polycrystalline Materials

KRYSINSKI Tomasz, MALBURET François
Mechanical Vibrations

KUNDU Tribikram
Advanced Ultrasonic Methods for Material and Structure Inspection

SIH George C. *et al.*
Particle and Continuum Aspects of Mesomechanics

Printed and bound by CPI Group (UK) Ltd, Croydon, CR0 4YY
13/09/2026
14961021-0004